WAYSIDE PLANTS OF THE ISLANDS

A Guide to the Lowland Flora of the Pacific Islands

including

HAWAI'I SAMOA TONGA TAHITI FIJI GUAM BELAU

by Dr. W. Arthur Whistler

Published by Isle Botanica
500 University Ave. # 1601
Honolulu, Hawaii 96826

Printed by Everbest Printing Company, Ltd., Hong Kong

Library of Congress no. 95-094213
ISBN: 0-9645426-0-9 (Soft Bound)
0-9645426-1-7 (Hard Cover)

This book is lovingly dedicated to Hazel Whistler Bullock, the matriarch of the Whistler clan, who has always been an inspiration to us all.

All photos in the text were taken by the author.

COVER: *Lantana camara* (Verbenaceae)

PREFACE

The basis for this book is an earlier one by the author, *Weed handbook of western Polynesia*, published in 1983. However, that book, which included photographs and descriptions of 82 common or troublesome weed species in western Polynesia, was not widely distributed. It was basically for farmers in Samoa and Tonga, and included a major section on weed control. The present book includes more than double the number (170) of species found in the earlier book, and is geared to the identification of the plants that grow naturally in the lowlands of the tropical Pacific islands (Polynesia, Micronesia, Melanesia).

For several years the author taught a lab class on ecology, which emphasized the common plants found in the lowlands of Hawai'i. The environment in Hawai'i has been greatly disturbed, and nowadays it is difficult to find a native plant in lowland areas. The same is true for the other islands of the tropical Pacific, where man has drastically changed the landscape and flora. The author has also spent many years doing botanical consulting in Hawai'i, Fiji, and Samoa, and to a lesser extent, in Micronesia. Based upon the field work and teaching experience, the most widespread, commonest, and/or most troublesome species of the lowlands of the tropical Pacific Islands have been selected for this book.

More than just the inclusion of new species was needed for this book; because of botanical research done over the last decade since the publication of *Weed handbook of western Polynesia* in 1983, many of the species names have been changed. The most up-to-date publications in the major island groups of the Pacific are the new flora of Fiji (Smith 1979—1991), a flora of Hawai'i (Wagner *et al.* 1990), and a checklist of the flora of Micronesia (Fosberg *et al.* 1979, 1982). In addition to these changes, all the photographs have been replaced by newer and hopefully better ones.

The book is most comprehensive for Hawai'i, which has the greatest number of naturalized lowland plants. Of the 170 species featured in the book, 153 are found there. The other groups listed in Table 1 are also well represented: Samoa with 120, Tonga with 106, the Society Islands with 121, Fiji with 121, Guam with 138, and Belau (formerly called Palau) with 102. In addition, another 85 species are mentioned in the text but are not shown in photographs.

Armed with this book, the botanist and lay person should be able to identify most of the naturalized plants in the lowlands of the islands. For the native plants that occur on the seashores of these regions, see *Flowers of the Pacific Island seashore* (Whistler 1992a). Hopefully, these books can help to foster an awareness of the fragile condition of the native vegetation, and highlight the need for measures to promote the protection of endangered species and areas of natural vegetation on tropical Pacific shores, as well as in native rain forests throughout the tropics.

Art Whistler
November 1994

ABOUT THE AUTHOR

Art Whistler was born near Death Valley, California, to which he attributes his early love of plants and vegetation. After receiving a B.A. and an M.A. at the University of California, he served three years with the U.S. Peace Corps in Western Samoa, where he taught high school biology. Resuming his schooling, he received a Ph.D. in Botany at the University of Hawai'i in 1979.

Since then he has made numerous research trips to Samoa, Tonga, Fiji, the Cook Islands, Tahiti, Guam, and elsewhere in the Pacific, working on the medicinal plants, ethnobotany, and flora of the islands. Currently he is a botanical consultant working on various projects in the Pacific Islands, and is an adjunct Associate Professor at the Department of Botany at the University of Hawai'i in Honolulu, where he has his research office. He has published several books on the botany of the Pacific Islands, including *Weed handbook of Western Polynesia* (1983), *Polynesian herbal medicine* (1992), *Tongan herbal medicine* (1992), and *Flowers of the Pacific island seashore* (1992), and has written numerous scientific articles on medicinal plants, ethnobotany, and floristics of Polynesia.

ACKNOWLEDGMENTS

I would like to thank the U.S. Army Corps of Engineers Operations Section, and especially Mike Lee, the Chief of the Operations Division, Honolulu District, for their financial support. Although I always have wanted to do this book, without their help it might never have become a reality. I am also grateful to Clyde Imada of the Bishop Museum Botany Department for his proofreading of the manuscript, Dr. Jacques Florence of O.R.S.T.O.M., for his comments on the plants in French Polynesia, and Dr. Lynne Raulerson, for her comments on Micronesia.

CONTENTS

INTRODUCTION

The Islands

The islands covered by this book fall into three geographic and cultural areas—Polynesia, Micronesia, and Melanesia. Within these areas are many archipelagoes of high volcanic and low coralline islands (atolls). However, for the purposes of this book, only the major ones are discussed and are included in Table 1. The naturalized and native lowland plants of the the remaining islands can be extrapolated since the weeds in one island or group of islands tend to be very similar to those of the adjacent ones.

Polynesia, meaning "many islands," is the huge, roughly triangle-shaped area between Hawai'i in the north, Easter Island in the southeast, and New Zealand in the southwest (see the front inside cover of this book). The largest group of islands within this triangle (with the exception of New Zealand, which has a temperate climate and is not included here) is Hawai'i. This archipelago, often called the "Hawaiian Islands" to distinguish it from the island of Hawai'i (commonly called the "Big Island"), consists of seven inhabited volcanic islands and several uninhabited ones, as well as a few atolls far to the northwest. The Big Island, with its twin peaks of Mauna Kea and Mauna Loa (both of which are over 4000 m in elevation), is the largest (10,466 km^2) and highest island in the region covered here. (Other Melanesian islands are larger, but are beyond the scope of this book.) Hawai'i probably has more naturalized species than the rest of the area combined, but many of these are temperate species restricted to higher elevation areas, and are not covered here.

The next largest archipelago in Polynesia is Samoa, which is situated in the South Pacific at about 14°S between Fiji, Tonga, and the Cook Islands. It is divided politically into the western portion called Western Samoa and the eastern portion called American Samoa. Western Samoa is an independent country comprising mainly the islands of Savai'i (1820 km^2) and 'Upolu (1110 km^2). American Samoa is much smaller (189 km^2), and comprises mainly Tutuila, Ta'ū , Ofu, and Olosega.

To the south of Samoa is the Tongan archipelago, which includes several hundred small islands, the largest of which is Tongatapu (250 km^2), where the capital, Nuku'alofa, is located. Over a thousand miles to the east is Tahiti. Tahiti is actually the largest of the Society Islands, which includes Mo'orea, Huahine, Taha'a, Ra'iatea, and Borabora. These islands, along with others in the area, including the sparsely inhabited Marquesas, are administered by France as "French Polynesia."

So Hawai'i, Samoa, Tonga, and Tahiti make up the Polynesian archipelagoes that are mentioned in this book and are shown in Table 1. The islands to the west comprise Melanesia, and are inhabited by a different cultural and racial group of people known as Melanesians. Only one Melanesian archipelago, Fiji, is listed in Table 1, and this is sometimes included in Polynesia since the culture there is more Polynesian than it is Melanesian. Fiji comprises hundreds of islands, some of them volcanic, but many of them composed of continental rock. The total area of Fiji is 18,235 km^2, which makes its slightly larger than Hawai'i. In fact, the main island of Fiji, Viti Levu, is almost exactly the same size as the Big Island of Hawai'i.

Micronesia comprises the area north of Melanesia and west of Polynesia, and nearly all of it is north of the equator (so it is not in the "South Pacific"). Most of the islands are relatively small, and most are atolls or low coralline islands. Two groups of islands are discussed and are included in Table 1, Guam and Belau. Guam, which is a U.S. territory, is part of the Marianas Islands, but is politically separate from the other islands of the archipelago, which are known as "The Commonwealth of the Northern Marianas." Most of Guam is comprised of raised limestone, and the total area is 559 km^2. Southwest of Guam is the Republic of Belau (formerly Palau), which comprises numerous islands and is also politically separate from the other islands in the region. Its islands have a total area of 311 km^2. Other major islands in the region are Saipan, Yap, Chuuk (Truk), Pohnpei (Ponape), and Kosrae (Kusiae); although these are not discussed in the text, their lowland flora, at least in the disturbed places, is very similar to that of the larger islands in the area—Guam and Belau.

The Lowlands

The area covered by this book is the "lowlands" of the islands, primarily because the lowland flora (the assemblage of plants) is very similar throughout the region. This was not always the case. Prior to the arrival of the first settlers in the Pacific Islands thousands of years ago, the lowlands were mostly covered with native lowland forest dominated by native species. Many of these species were endemic to one island or group—i.e., they were restricted to that group. It is estimated that about 89% (Wagner *et al.* 1990) of the native flora of Hawai'i was endemic to Hawai'i. This high rate can be attributed to the distinct isolation of that archipelago—the nearest high islands or continents are thousands of miles away.

However, with the arrival of man, the lowland landscape underwent a drastic change. The settlers cut down the forests to plant their crops and to build their villages. They also introduced non-native plants that had the ability to out-compete the native plants in disturbed environments. The result is today, after thousands of years of occupation by man, and particularly by Europeans in the last two centuries, there is very little native lowland forest left. The lowlands are so disturbed that the "alien" (introduced) or invader plant species have nearly totally replaced the native species. In some places, such as Hawai'i, it is difficult to find native plants in the lowlands, and you have to go up into the mountains, which are relatively less disturbed, before you can find significant number of native species.

So this is the current landscape in the lowlands—very disturbed plant communities dominated by introduced species.

The Lowland Flora

The title of this book calls the plants included here "wayside plants," for want of a better term, and implies that these plants are found along trails and tracks. It could have instead been entitled "Lowland Plants of the Islands," since it basically covers the lowland flora of the islands, or "Weeds of the Islands," since the vast majority of plants now found in the lowlands of the Pacific Islands are introduced species or "weeds."

A "weed" may be defined as any plant growing where it is not wanted. This definition is based on the location where the plant is growing or on the economic importance attributed to it by man, rather than on the species itself. In other words, a plant growing in one place may be considered to be a weed, but when growing in another, it is not. For example, *Brachiaria mutica* is a valuable forage grass when it occurs in pastures, but is considered a serious weed on roadsides and in irrigation ditches. Likewise, plants such as *Mimosa pudica* and *Lantana camara* are considered to be desirable ornamentals in some parts of the world, but are serious pasture weeds in the tropics. On the other hand, some species, particularly poisonous ones, are considered to be weeds wherever they occur. In actuality, there is no hard and fast line between weeds and non-weeds, and borderline species (like some ornamentals) that are sometimes found growing away from where they were planted are put into an intermediate category, "casual adventive."

There are a number of characteristics that make these introduced species so successful. In general, the successful invaders are more efficient in their use of water and minerals, and simply out-compete native species. Some of the other notable characteristics of successful invaders are listed below.

1. Production of numerous seeds— This gives them a competitive advantage in reproduction, enabling them to simply overwhelm native species or crop plants by sheer numbers. Also, with greater numbers of seeds, they are more likely to spread to other areas.

2. Dispersability— Most of the successful invaders have seeds that are readily dispersed. This is an important characteristic because those with poor means of dispersal are slower to spread. Some of these invaders, particularly members of the sunflower family (Asteraceae), have seeds with silky, parachute-like bristles or plumes that allow them to be dispersed long distances by the wind. Some of the fastest spreading species have fruits or seeds that bear a bristle or "awn" (like *Bidens alba*, beggar's tick), are spiny or barbed (*Cenchrus echinatus*, sand bur), or are fuzzy or hairy (like *Desmodium incanum*). These fruits and seeds readily stick to clothing, feathers, or fur and are thereby dispersed. Another common means that some of these species employ to spread their seeds is a fleshy, colored fruit containing seeds that can pass unharmed through a bird's digestive tract. Other easily dispersed species have sticky seeds that adhere to the outside of birds, animals, or humans, or in mud stuck to shoes.

3. Lack of biological controls— The invader species are often kept under control in their native habitat by insects and/or specific plant diseases, but when they are dispersed to the Pacific Islands and elsewhere, these controlling agents are left far behind. This lack of biological control makes the invaders more aggressive and successful in their new island home.

4. Vegetative reproduction— Many of the invaders have the ability to spread by vegetative means as well as (or instead of) by seeds. One of the world's worst weeds, *Cyperus rotundus*, forms tubers at the ends of the underground stems, and can spread from field to field on improperly cleaned plows. *Clerodendrum chinense*, a serious pest in Samoa, uses only asexual means (root suckers) to spread because it is sterile and cannot produce seeds.

Consequently, the vast majority of plants included in the following pages are non-native species, most of which would be classified as "weeds." This leads to a homogeneity of habitats in the Pacific that were formerly inhabited by a diversity of native species. Thus, the lowland flora of the Pacific islands is basically a flora of the most easily dispersed weeds of the tropical regions of the world.

HOW TO USE THIS BOOK

The species in this book are listed by scientific (or "Latin") name in capital letters across the top of each page. All scientific names consist of two parts—the generic and the specific name. The genus name is capitalized and the specific name is lower-cased, and both are italicized (since they are in Latin). Genera (the plural of genus) usually contain a number of species. For example, *Bidens alba* is in the genus *Bidens*, as is another species mentioned in the text, *Bidens pilosa*. Scientific names are important because they are universal, are used by scientists in every country, and are coined according to a fixed set of rules that ensures their accuracy. Common names, however, often vary from country to country, are often misleading, and frequently are inconsistent even in the same country. Moreover, most naturalized or weedy species, such as the ones included in this book, lack common names.

Under the scientific name is the name of the "family," a taxonomic grouping of related genera. All family names end in -aceae (although formerly some, such as Compositae, now called Asteraceae, did not). The family is also identified by a common name, to give the layman an idea of what is in the group, such as "Sunflower Family" (Asteraceae) for the family that includes sunflowers and asters.

Under the family name are the common names. These are usually in English; if not, their origin is noted in parentheses. Below the common name is the distribution of the species within the Pacific Islands, or more specifically, within the major island groups—Hawai'i, Samoa, Tonga, Tahiti (actually the Society Islands), Fiji, Guam, and Belau.

Beside the photograph of the plant is a botanical description. This is rather technical in nature and is more for the scientist than the layman. However, a glossary containing the terms is found at the back of the book. In the botanical descriptions the major categories—leaves, inflorescence, calyx, corolla, and fruit—are printed in bold. The measurements are given in metric units, since this is the system used in nearly all the world except the United States. A metric scale in centimeters is found on the back outside cover of the book. One inch equals 2.54 cm; one meter equals 39.36 inches.

The first paragraph below the photo includes several things. It begins with the scientific name with the author citations included. These citations are sometimes needed by scientists for reasons of clarity; there has been much confusion in the use of scientific names, and by adding the author citation, it indicates that this is the plant recognized by that particular author. Sometimes there are two author names after the species, one of which is in parentheses. For example, the plant *Centella asiatica* (p. 27) is followed by "(L.) Urb." This indicates that the plant was originally named by L. (Linnaeus) in a different genus (in this case, *Hydrocotyle)*, but at a later date a botanist named Urban decided that it didn't belong in that genus and transferred it to *Centella*. Author citations are usually resorted to only in taxonomic publications, and are of little consequence to the layman, but are included here for those who may need them.

Following the author citation in the first paragraph is the origin and/or the native range of the species. As noted earlier, only a few of the species included in this book are native to the Pacific Islands, the rest being "alien" or "introduced" species. Some of the introduced species were brought by the original inhabitants prior to the European discovery of the islands. These are described as ancient introductions (also often called Polynesian introductions in Polynesia, or aboriginal introductions). For the majority of the species—the introduced ones—the date and location of the record is listed. These are often somewhat misleading, since accurate botanical records were not kept until recently. They are based almost entirely on Hawai'i, Samoa, and Fiji, which have the most readily accessible records.

Next in the paragraph is the habitat, i.e., the types of areas in which the plant is found, and its abundance or frequency within the Pacific Islands. Elevational range, when available, is also recorded. Nearly all of the species included in the book can be found near sea level, but the highest elevations at which they are found varies from species to species.

The last paragraph is a diagnostic description that includes the most salient features to aid in identification. The terminology in this description is somewhat simpler than the botanical description, but non-botanists may still need to resort to the glossary at the back of the book. Following this description are synonyms—names other than the correct one—that have been used in the past for the plant. And finally, related species are noted in many cases.

Another character not discussed on each page, but which is found in the index, is the wetland status of the species. One of the most important natural resources of the world is its wetlands. These, however, are fast disappearing, and in order to preserve wetlands, recent U.S. legislation has defined wetlands based partly on the species present. Species that are never found in wetlands are often unclassified by this system, or are designated as "upland" species; species that are usually found in dry soil, but which are sometimes in wetlands are called "facultative upland species" (FACU); species that are found nearly equally in wet or dry soil are called "facultative species" (FAC); species that are usually found in wetlands are called "facultative wetland species" (FACW); and species virtually always associated with wetlands are called "obligate wetland species" (OBL). These designations are included in the index, based on Reed (1988), as an aid to biologists who need to determine wetland boundaries based on the species present. Further information on wetlands is found in *A guide to Pacific wetland plants* (Stemmermann 1981).

THE ARRANGEMENT OF SPECIES

The arrangement of species in a book like this is always a problem. The ideal way to arrange them is so that everyone can easily find what they are looking for. However, such a goal is unattainable, and an arrangement that is easiest for most people is perhaps the ideal to be attained. There are several basic ways that the species could be arranged.

(1) By flower color— This has the advantage of a visual cue that is understood by everybody. The disadvantage is that sometimes the color will vary within a species, and two species that are closely related but different in flower color will not be arranged together. Another disadvantage is that trees, vines, and herbs, which are different in life form and which may be totally unrelated taxonomically, will be all mixed together.

(2) By life form—In this arrangement, the species are placed in categories such as tree, shrub, herb, vine, and grass/sedge. If you know the life form, you go to the appropriate section and look through it until you find the species. The disadvantages of this are that species don't always fall neatly into these five artificial categories, and there still may be dozens of plants to look through.

(3) By taxonomic order— Botanists who study the evolution of plants—taxonomists—often arrange species in an order based on their relationship to each other. By custom, the nettle family (Urticaceae) is placed near the front of the dicots (members of one of the two main divisions of flowering plants, the other being monocots), and the sunflower family (Asteraceae or Compositae) at the back of the dicot section. This is fine for taxonomists who know the arrangement, but impossible for those who don't.

(4) By alphabetical order— Arranging species in alphabetical order is one simple way to do it, but the disadvantages are that the scientific names are sometimes revised and changed, and species that are related to each other will not be found together. Also, most people who use a book such as this one do not know the scientific names anyway.

(5) By a combination of methods— The species may be arranged into the life form categories first (as is done in *Flowers of the Pacific island seashore*) and then alphabetically; they may be arranged in alphabetical order by family and then in alphabetical order by species within the family (as is done in Table 1); or they may be arranged by life form category and then by flower color within the categories.

After weighing the pros and cons, the author decided that the plants would best be arranged by family in alphabetical order, and by species in alphabetical order within the families . Thus all the plants of a family are together, and if you know the family, you can easily find them in their alphabetical arrangement.

TABLE 1. THE SPECIES COVERED

The species in **bold** are featured in the text; those indented and in regular type are discussed under the featured species above it. "X" indicates the species is present; "R" indicates it is present but rare; and "C" indicates it is present only in cultivation.

FAMILY Species	HAW.	SAM.	TON.	SOC.	FIJI	GUAM	BEL.
DICOTS							
ACANTHACEAE							
Asystasia gangetica	X	R	X	X	X	X	—
Blechum pyramidatum	X	X	X	—	X	X	X
Justicia procumbens	—	X	—	—	—	—	—
Ruellia prostrata	R	X	—	—	—	—	—
Ruellia tuberosa	—	X	—	—	—	—	—
AMARANTHACEAE							
Achyranthes aspera	X	X	X	X	X	X	X
Alternanthera pungens	X	—	—	—	—	—	—
Alternanthera caracasana	X	—	—	—	—	—	—
Alternanthera sessilis	X	X	X	X	X	X	X
Amaranthus spinosus	X	X	—	—	X	X	X
Amaranthus dubius	X	—	—	X	X	—	—
Amaranthus hybridus	X	—	—	—	X	—	
Amaranthus viridis	X	X	X	X	X	X	X
Amaranthus lividus	X	X	X	R	—	—	—
ANACARDIACEAE							
Schinus terebinthifolius	X	C	—	C	C	C	—
APIACEAE							
Centella asiatica	X	X	X	X	X	X	X
ARALIACEAE							
Schefflera actinophylla	X	—	—	X	X	C	—
ASTERACEAE							
Ageratum conyzoides	X	X	X	X	X	X	X
Ageratum houstonianum	X	—	—	X	X	—	—

FAMILY Species	HAW.	SAM.	TON.	SOC.	FIJI	GUAM	BEL.
Bidens alba	X	X	—	—	—	X	X
Bidens cynapiifolia	X	—	—	—	—	—	—
Bidens pilosa	X	X	X	X	X	X	X
Calyptocarpus vialis	X	—	—	—	—	X	—
Chromolaena odorata	—	—	—	—	—	X	X
Conyza bonariensis	X	R	X	X	X	X	X
Conyza canadensis	X	—	X	—	X	X	X
Crassocephalum crepidioides	X	X	X	X	X	X	X
Erechtites valerianifolia	X	X	X	—	X	—	—
Elephantopus mollis	X	—	X	X	X	X	—
Emilia fosbergii	X	—	—	X	X	X	—
Emilia sonchifolia	X	X	X	X	X	X	X
Mikania micrantha	—	X	R	X	X	—	—
Mikania scandens	—	—	—	—	—	X	—
Pluchea carolinensis	X	—	—	R	—	X	X
Pluchea indica	X	R	—	—	—	X	X
Pluchea x fosbergii	X	—	—	—	—	X	—
Pseudelephantopus spicatus	X	X	X	X	X	R	—
Sonchus oleraceus	X	—	X	X	X	X	—
Synedrella nodiflora	X	X	X	X	X	X	X
Eleutheranthera ruderalis	—	X	—	—	X	—	—
Tridax procumbens	X	X	X	X	X	X	X
Verbesina encelioides	X	—	—	—	—	—	—
Vernonia cinerea	X	X	X	X	X	X	X
Wedelia trilobata	X	X	—	X	—	X	X
Xanthium strumarium	X	—	—	X	—	—	—
Xanthium pungens	—	—	—	—	X	—	—
Youngia japonica	X	X	X	X	X	X	—
BIGNONIACEAE							
Spathodea campanulata	X	X	C	X	X	X	C
BORAGINACEAE							
Heliotropium procumbens	X	X	—	—	—	X	—
Heliotropium amplexicaule	X	—	—	—	X	—	—
CACTACEAE							
Opuntia ficus-indica	X	—	—	—	—	—	—
Opuntia cochenillifera	X	—	—	C	—	X	X
CHENOPODIACEAE							
Atriplex semibaccata	X	—	—	—	—	—	—
Atriplex suberecta	X	—	—	—	—	—	

FAMILY Species	HAW.	SAM.	TON.	SOC.	FIJI	GUAM	BEL.
Chenopodium murale	X	—	R	—	—	—	—
Chenopodium ambrosioides	X	—	—	—	X	X	—
Chenopodium carinatum	X	—	—	—	—	—	—
CONVOLVULACEAE							
Ipomoea alba	X	R	R	X	X	X	—
Ipomoea cairica	X	—	X	—	X	—	—
Ipomoea indica	X	R	X	X	X	X	X
Ipomoea littoralis	X	X	X	X	X	X	X
Ipomoea obscura	X	—	—	X	X	X	—
Ipomoea triloba	X	—	—	—	—	X	X
Merremia aegyptia	X	—	—	—	—	X	—
Merremia peltata	—	X	X	X	X	X	X
Merremia umbellata	X	X	—	X	X	—	—
Merremia tuberosa	X	C	—	X	X	C	C
Operculina ventricosa	—	X	X	—	—	X	—
Operculina turpethum	—	X	R	X	X	—	X
Stictocardia tiliifolia	X	X	X	X	X	X	X
CRASSULACEAE							
Kalanchoë pinnata	X	R	R	X	X	R	X
Kalanchoë tubiflora	X	—	—	—	—	C	—
CUCURBITACEAE							
Coccinia grandis	X	—	X	—	X	X	—
Momordica charantia	X	X	X	X	X	X	X
EUPHORBIACEAE							
Acalypha lanceolata	—	X	X	—	X	—	X
Acalypha indica	—	X	—	—	—	X	X
Chamaesyce hirta	X	X	X	X	X	X	X
Chamaesyce hypericifolia	X	X	X	X	X	X	X
Chamaesyce hyssopifolia	X	X	—	—	X	—	—
Chamaesyce prostrata	X	X	X	X	X	X	X
Chamaesyce thymifolia	X	X	X	X	—	X	X
Euphorbia heterophylla	X	R	—	—	—	X	—
Euphorbia cyathophora	X	X	X	X	X	X	X
Phyllanthus amarus	—	X	X	X	X	X	X
Phyllanthus urinaria	R	X	—	X	X	R	X
Phyllanthus debilis	X	X	—	X	X	X	X
Phyllanthus tenellus	X	—	—	X	—	—	—
Ricinus communis	X	X	X	X	X	X	X
FABACEAE							
Acacia farnesiana	X	—	—	X	X	X	—

FAMILY Species	HAW.	SAM.	TON.	SOC.	FIJI	GUAM	BEL.
Alysicarpus vaginalis	X	X	X	X	X	X	X
Calopogonium mucunoides	—	X	—	X	X	X	X
Chamaecrista nictitans	X	R	X	—	X	X	X
Chamaecrista mimosoides	—	—	—	—	X	X	X
Crotalaria pallida	X	X	X	X	X	X	X
Crotalaria incana	X	X	R	—	X	—	—
Desmanthus virgatus	X	R	—	X	X	X	X
Desmodium incanum	X	X	X	X	X	X	X
Desmodium sandwicense	X	—	—	—	—	—	—
Desmodium tortuosum	X	X	X	—	X	R	—
Desmodium triflorum	X	X	X	X	X	X	X
Desmodium heterophyllum	—	X	X	—	X	X	—
Indigofera suffruticosa	X	X	X	X	X	X	—
Indigofera spicata	X	—	—	X	X	—	—
Leucaena leucocephala	X	X	X	X	X	X	X
Macroptilium lathyroides	X	X	X	X	X	X	—
Macroptilium atropurpureum	X	X	X	X	X	X	—
Mimosa invisa	X	—	X	X	—	X	—
Mimosa pudica	X	X	X	X	X	X	X
Pithecellobium dulce	X	—	—	C	C	C	C
Prosopis pallida	X	—	—	—	C	C	—
Pueraria lobata	X	X	X	C	X	X	X
Samanea saman	X	X	C	X	C	C	C
Senna occidentalis	X	X	X	X	X	X	X
Senna tora	—	X	X	X	X	X	—
Senna surattensis	X	—	—	C	—	C	—
Senna pendula	X	—	—	—	—	—	—
LAMIACEAE							
Hyptis pectinata	X	X	X	X	X	X	X
Hyptis suaveolens	X	—	—	—	—	X	—
Hyptis rhomboidea	—	X	—	X	—	X	X
Leonotis nepetifolia	X	—	—	—	—	—	—
Ocimum gratissimum	X	X	—	X	—	—	—
Ocimum basilicum	X	X	X	X	X	X	—
Ocimum tenuiflorum	—	X	—	—	X	X	X
LYTHRACEAE							
Cuphea carthagenensis	X	X	X	X	X	—	—
MALVACEAE							
Abutilon grandifolium	X	—	—	X	—	—	—
Abutilon indicum	R	—	—	R	—	X	—

FAMILY Species	HAW.	SAM.	TON.	SOC.	FIJI	GUAM	BEL.
Malva parviflora	X	—	—	—	—	—	—
Malvastrum coromandelianum	X	R	X	X	X	X	—
Sida fallax	X	—	—	R	—	—	X
Sida cordifolia	X	—	R	R	—	X	X
Sida rhombifolia	X	X	X	X	X	X	X
Sida acuta	X	X	X	X	X	X	X
Sida spinosa	X	—	—	X	—	—	X
Urena lobata	X	X	X	X	X	X	X
MELASTOMATACEAE							
Clidemia hirta	X	X	—	—	X	—	X
MORACEAE							
Ficus microcarpa	X	—	—	C	—	X	X
MYRTACEAE							
Psidium cattleianum	X	C	—	X	X	—	X
Psidium guajava	X	X	X	X	X	X	X
Syzygium cumini	X	C	C	X	X	C	C
NYCTAGINACEAE							
Boerhavia coccinea	X	—	—	—	—	X	—
Boerhavia repens	X	X	X	X	X	X	X
ONAGRACEAE							
Ludwigia octovalvis	X	X	X	X	X	X	X
Ludwigia hyssopifolia	—	X	—	—	X	X	X
OXALIDACEAE							
Oxalis corniculata	X	X	X	X	X	X	X
Oxalis barrelieri	—	X	—	R	X	—	X
Oxalis corymbosa	X	X	—	X	X	—	—
PASSIFLORACEAE							
Passiflora foetida	X	X	X	X	X	X	X
Passiflora edulis	X	C	C	C	X	C	—
Passiflora suberosa	X	R	—	X	X	X	—
PHYTOLACCACEAE							
Rivina humilis	X	—	X	X	X	—	—
PLANTAGINACEAE							
Plantago major	X	X	X	X	X	R	—
Plantago debilis	X	—	X	—	—	—	—
Plantago lanceolata	X	R	X	X	—	—	—
POLYGALACEAE							
Polygala paniculata	X	X	—	R	X	X	X
PORTULACACEAE							
Portulaca oleracea	X	X	X	X	X	X	X

FAMILY Species	HAW.	SAM.	TON.	SOC.	FIJI	GUAM	BEL.
Portulaca pilosa	X	—	—	—	—	—	—
PROTEACEAE							
Grevillea robusta	X	—	C	C	—	C	—
ROSACEAE							
Rubus rosifolius	X	—	—	X	—	—	—
Rubus argutus	X	—	—	—	—	—	—
RUBIACEAE							
Spermacoce assurgens	X	X	X	X	X	X	X
Spermacoce bartlingiana	—	X	—	—	X	—	X
Spermacoce mauritiana	X	X	—	—	X	—	—
RUTACEAE							
Triphasia trifolia	—	R	—	—	X	X	—
SOLANACEAE							
Nicandra physalodes	X	—	—	—	—	—	—
Nicotiana glauca	X	—	—	—	—	—	—
Nicotiana tabacum	X	C	C	C	X	C	C
Physalis angulata	X	X	X	X	X	X	X
Physalis peruviana	X	R	R	R	X	X	—
Solanum americanum	X	X	X	X	X	X	X
Solanum linnaeanum	X	—	—	—	X	—	—
Solanum seaforthianum	X	—	—	—	—	—	—
Solanum torvum	X	X	X	X	X	X	X
Solanum mauritianum	X	—	X	—	X	—	—
STERCULIACEAE							
Waltheria indica	X	R	R	X	X	X	—
TILIACEAE							
Triumfetta rhomboidea	X	X	X	X	X	X	X
Triumfetta semitriloba	X	R	—	—	—	X	—
URTICACEAE							
Laportea interrupta	—	X	X	X	X	X	X
Laportea ruderalis	—	—	—	X	—	X	X
Pilea microphylla	X	X	X	X	X	X	X
Pilea peploides	X	—	—	—	—	—	—
VERBENACEAE							
Clerodendrum chinense	X	X	—	X	X	—	—
Lantana camara	X	X	X	X	X	X	X
Stachytarpheta jamaicensis	X	X	R	—	—	X	X
Stachytarpheta dichotoma	X	—	—	—	—	—	—
Stachytarpheta urticifolia	X	X	X	X	X	X	X
Stachytarpheta cayennensis	—	X	—	X	—	—	—

FAMILY Species	HAW.	SAM.	TON.	SOC.	FIJI	GUAM	BEL.
Verbena litoralis	X	—	—	—	—	—	—
Verbena bonariensis	X	—	—	—	X	—	—
Vitex parviflora	—	—	—	—	—	X	X

MONOCOTS

FAMILY Species	HAW.	SAM.	TON.	SOC.	FIJI	GUAM	BEL.
AGAVACEAE							
Agave sisalana	X	—	—	C	X	C	C
Furcraea foetida	X	—	—	C	X	C	—
COMMELINACEAE							
Commelina benghalensis	X	X	X	—	—	X	—
Commelina diffusa	X	X	X	X	X	X	X
CYPERACEAE							
Cyperus compressus	X	X	X	X	X	X	X
Cyperus gracilis	X	—	—	—	—	—	—
Cyperus rotundus	X	X	X	X	X	X	X
Fimbristylis dichotoma	X	X	X	X	X	X	X
Kyllinga nemoralis	X	X	X	X	X	X	X
Kyllinga brevifolia	X	X	X	X	X	X	X
Kyllinga polyphylla	—	X	—	X	X	—	—
Pycreus polystachyos	X	X	X	X	X	X	X
DIOSCOREACEAE							
Dioscorea bulbifera	X	X	X	X	X	X	X
POACEAE							
Axonopus compressus	X	X	X	X	X	X	X
Axonopus fissifolius	X	—	—	X	X	—	—
Bothriochloa pertusa	X	—	—	—	—	X	—
Bothriochloa bladhii	—	X	X	X	X	X	X
Brachiaria mutica	X	X	X	X	X	X	X
Brachiaria subquadripara	X	X	X	—	X	X	X
Brachiaria paspaloides	—	X	X	X	X	X	X
Cenchrus ciliaris	X	—	—	X	X	—	—
Cenchrus echinatus	X	X	X	X	X	X	X
Chloris barbata	X	X	X	X	X	X	X
Chloris virgata	X	—	—	R	—	—	—
Chloris radiata	X	R	R	—	—	X	X
Chloris divaricata	X	—	X	—	X	—	—
Chrysopogon aciculatus	X	X	X	X	X	X	X
Cynodon dactylon	X	X	X	X	X	X	X
Dactyloctenium aegyptium	X	X	X	X	X	X	X

FAMILY Species	HAW.	SAM.	TON.	SOC.	FIJI	GUAM	BEL.
Digitaria ciliaris	X	X	X	X	X	X	X
Digitaria horizontalis	—	X	—	X	—	—	—
Digitaria radicosa	X	X	X	X	X	X	X
Digitaria insularis	X	—	—	—	—	X	—
Digitaria setigera	X	X	X	X	X	X	X
Digitaria violascens	X	X	X	X	X	X	X
Echinochloa colona	X	X	X	X	X	X	X
Echinochloa crus-galli	X	—	—	X	X	—	X
Echinochloa stagnina	—	X	—	—	X	—	—
Eleusine indica	X	X	X	X	X	X	X
Eragrostis tenella	X	X	X	X	X	X	X
Eragrostis cilianensis	X	—	—	—	—	—	—
Eragrostis pectinacea	X	—	—	—	—	—	—
Eriochloa procera	—	X	X	—	X	X	—
Eustachys petraea	—	—	—	—	—	X	X
Heteropogon contortus	X	R	X	—	X	X	—
Melinus minutiflora	X	R	X	X	X	X	X
Oplismenus compositus	X	X	X	X	X	X	X
Oplismenus hirtellus	X	X	X	X	X	X	X
Panicum maximum	X	X	X	X	X	X	X
Panicum repens	X	—	—	—	—	—	X
Paspalum conjugatum	X	X	X	X	X	X	X
Paspalum fimbriatum	X	R	X	—	—	X	—
Paspalum dilatatum	X	X	X	X	X	X	—
Paspalum paniculatum	—	X	X	X	X	X	—
Paspalum urvillei	X	X	—	—	X	X	—
Paspalum setaceum	—	X	—	—	—	X	—
Paspalum orbiculare	X	X	X	X	X	X	X
Pennisetum polystachion	X	—	—	X	X	X	X
Pennisetum purpureum	X	X	—	X	X	X	X
Pennisetum setaceum	X	—	—	—	X	—	—
Rhynchelytrum repens	X	—	X	X	X	X	—
Saccharum spontaneum	X	—	—	—	—	X	X
Setaria verticillata	X	—	—	X	—	X	—
Setaria pumila	—	X	X	X	X	X	X
Setaria parviflora	X	—	—	—	—	—	—
Sorghum sudanense	X	X	X	X	—	X	—
Sorghum halepense	X	—	—	—	X	X	X
Sporobolus diander	X	X	X	X	X	X	X
Sporobolus fertilis	X	—	X	X	X	X	X

HAW. = Hawai'i; SAM. = Samoa; TON.= Tonga; SOC.= Society Islands; BEL. = Belau.

ASYSTASIA GANGETICA
Acanthaceae (Acanthus Family)

COMMON NAMES: Chinese violet
DISTRIBUTION: Hawai'i, Samoa (rare), Tonga, Tahiti, Fiji, Guam

Weak-stemmed perennial herb. Stems up to 1.5 m or more in length, appressed-pubescent. **Leaves** opposite, simple, blade ovate, 2.5—8 x 1.5—4 cm, acute at the tip, mostly rounded to subcordate at the base; surfaces sparingly pubescent; margins entire; petiole 0.5—4 cm long. **Inflorescence** a few-flowered, 1-sided raceme 4—8 cm long. **Calyx** 6—9 mm long, deeply divided into 5 linear lobes, subtended by a linear to triangular bract 1—2 mm long; pedicel 1—2 mm long. **Corolla** funnelform, 3—4 cm long, somewhat 2-lipped, pale lavender to purple or white, shallowly divided into 5 spreading, rounded lobes. Stamens 4, in 2 pairs, epipetalous. Ovary superior, 2-celled. **Fruit** a 2-lobed, clavate capsule 2—3 cm long with a short beak, 4-seeded.

Asystasia gangetica (L.) T. Anders. is native from Africa to Malaysia and was first recorded from the Pacific Islands in 1925 (Hawai'i). Although common to abundant in disturbed, lowland urban areas and disturbed *Prosopis* and *Leucaena* forests of Hawai'i, it is less frequent in most of the other Pacific archipelagoes. It is sometimes cultivated as an ornamental because of its showy flowers.

This herb can be distinguished by its weak stems, opposite leaves, one-sided, several-flowered racemes, showy, funnel-shaped, purple, lavender, or white corollas with 5 rounded lobes, and 2-lobed, club-shaped capsules. Synonym: *Asystasia coromandelianum* Nees.

BLECHUM PYRAMIDATUM
Acanthaceae (Acanthus Family)

COMMON NAMES: none
DISTRIBUTION: all the main island groups except Tahiti

Prostrate to erect herb 10—60 cm in height. Stems pubescent, rooting at the lower nodes. **Leaves** opposite, simple, blade ovate to elliptic, 2—8 x 1—4 cm, acute to attenuate at the apex, attenuate at the base; surfaces sparingly appressed-pubescent to glabrous; margins wavy; petiole 0.5—2 cm long. **Inflorescence** a terminal spike 2—6 cm long, with the flowers concealed by overlapping, pubescent, ovate, leafy bracts 1—2.5 cm long. **Calyx** 2.5—4 mm long, divided over halfway into 5 linear lobes; pedicel short. **Corolla** funnel-shaped, 10—14 mm long, white, divided into 5 round lobes. Stamens in 2 pairs, included. Ovary superior, 2-celled. **Fruit** a brown, spindle-shaped capsule 3—7 mm long, splitting into 2 spreading valves, each with 4—6 discoid seeds 1.2—2 mm long.

Blechum pyramidatum (Lam.) Urb. is native to tropical America and was first recorded from the Pacific Islands in 1929 (Fiji). It is common to locally abundant (especially in Fiji and Samoa) in relatively wet, disturbed places such as pastures, roadsides, marshland boundaries, and croplands up to 600 m elevation, and is occasional in forest clearings.

This prostrate to erect herb can be distinguished by its opposite, ovate leaves, terminal racemes densely covered with ovate, leafy bracts, small, white, funnel-shaped flowers, and small spindle-shaped capsules containing 8—12 disc-shaped seeds. Synonym: *Blechum brownei* Juss., which is the name used in the flora of Hawai'i (Wagner *et al.* 1990), but corrected in Smith (1991).

JUSTICIA PROCUMBENS
Acanthaceae (Acanthus Family)

COMMON NAMES: none
DISTRIBUTION: Samoa

Decumbent perennial herb up to 40 cm in height. Stems pubescent, rooting at the lower nodes. **Leaves** opposite, simple, elliptic to ovate, 1—4 x 0.6—1.5 cm, acute at the apex and base; surfaces glabrous to sparsely pubescent; margins ciliate; petiole 1—4 mm long. **Inflorescence** of 1—3 dense, ovoid to cylindrical, axillary or terminal racemes 0.6—6 cm long. **Calyx** 3—6 mm long, deeply divided into 5 linear lobes with hairy margins, subsessile, subtended by two bracts similar to the calyx lobes. **Corolla** bilabiate, 4—7 mm long, shallowly divided into 2 lobes on the small upper lip and 3 on the lower, larger, suborbicular lip, lavender with white markings. Stamens 2, epipetalous, included. Ovary superior, 2-celled. **Fruit** an ovoid capsule 4—5 mm long, splitting and spreading open by 2-valves, 4-seeded.

Justicia procumbens L. is native to tropical America, and although first recorded in Samoa in 1955, it is absent from the other main Pacific Islands so far. It is common to locally abundant in disturbed lowland places such as roadsides and lawns on all the main islands of American and Western Samoa.

This low herb can be distinguished by its stems that root at the lower nodes, small, elliptic, opposite leaves, dense, ovoid or cylindrical, spike-like racemes, linear, hairy bracts, tiny, lavender, 2-lipped flowers, and small, ovoid, 4-seeded capsules.

RUELLIA PROSTRATA
Acanthaceae (Acanthus Family)

COMMON NAMES: vao uli (Samoa)
DISTRIBUTION: Hawai'i (rare), Samoa

Erect to ascending herb 15—50 cm in height. Stems sparingly pubescent, rooting at the lower nodes. **Leaves** opposite, simple, blade ovate to elliptic, 1.5—7 x 0.8—3.5 cm, acute at the apex and base; surfaces sparingly appressed-pubescent, margins subentire; petiole 0.8—4 cm long. **Inflorescence** of solitary, subsessile, axillary flowers. **Calyx** 6—12 mm long, divided to near the base into 5 linear lobes, subtended by 2 obovate bracts 1—2.5 cm long. **Corolla** funnelform, 22—30 mm long, lavender, with 5 rounded lobes. Stamens in 2 pairs, included. Ovary superior, 2-celled. **Fruit** a densely puberulent, clavate capsule 15—22 mm long, acute-tipped, splitting at maturity into 2 spreading valves, each with 5—8 flattened-ovoid seeds.

Ruellia prostrata Poiret is native to Java, Indonesia and was first recorded from the Pacific Islands in 1944 (Hawai'i). It is common to abundant in Samoa in lowland disturbed areas such as under crop trees (e.g., cacao) and along roadsides. In the rest of the Pacific Islands, it is recorded only from Hawai'i, where it is uncommon.

This erect to ascending herb can be distinguished by its opposite, ovate leaves distinctly lighter on the lower surface, solitary, axillary, lavender, funnel-shaped flowers subtended by a pair of ovate bracts, and a club-shaped capsule splitting open by two valves. Synonym: *Ruellia repens* of Hawaiian authors. A similar weedy, but less frequent, species in Samoa, *Ruellia tuberosa* L., has much larger flowers and subwoody stems.

ACHYRANTHES ASPERA
Amaranthaceae (Amaranth Family)

COMMON NAMES: tamatama (Samoa, Tonga); aerofai (Tahiti)
DISTRIBUTION: all the main island groups

Sparsely branching subshrub herb up to 1.5 m in height. Stems somewhat angled and grooved, appressed-pubescent. **Leaves** opposite, simple, blade obovate to elliptic, 2—12 x 1—7 cm, acute to rounded and shortly acuminate at the apex, acute to acuminate at the base; surfaces pubescent; margins entire; petiole 0.5—3 cm long. **Inflorescence** a terminal spike 10—40 cm long, bearing reflexed, green, spikelet-like flowers crowded at the tip when young; bracts ovate, membranous, 1.5—2.5 mm long, with a spine of similar length. **Calyx** of 5 subequal lanceolate, green sepals 4—7 mm long, subtended by 2 stiff bracteoles otherwise similar to the bract. **Corolla** absent. Stamens 5, free. Ovary superior. **Fruit** an obovoid, indehiscent, 1-seeded utricle 2—3 mm long, truncate at the top.

Achyranthes aspera L. is native or was perhaps an ancient introduction to the Pacific Islands (but a European introduction to Hawai'i) and ranges from Southeast Asia to Polynesia. It is occasional to uncommon in native coastal areas in most of the Pacific, but is locally common in disturbed forests in Hawai'i. The grass-like fruits readily stick to clothes, feathers, and fur. Some medicinal uses are reported from Polynesia (Whistler 1992b).

This subshrub can be distinguished by its pubescent stems and foliage, opposite, mostly obovate leaves, long terminal spikes bearing flowers bent back and looking like grass spikelets, 5 lanceolate sepals 4—7 mm long, and absence of a corolla. Synonym: *Achyranthes indica* (L.) Mill. Several endemic species of the genus occur in Hawai'i, some of them quite rare.

ALTERNANTHERA PUNGENS
Amaranthaceae (Amaranth Family)

COMMON NAMES: khaki weed
DISTRIBUTION: Hawai'i

Perennial prostrate herb. Stems 10—50 cm long, pubescent, often rooting at the nodes. **Leaves** opposite, simple, blade ovate to obovate, mostly 1—3 x 0.8—1.5 cm, subround to acute and mucronate at the apex, cuneate at the base; surfaces nearly glabrous; margins entire; petiole 2—8 mm long. **Inflorescence** a sessile, ovoid, green to straw-colored spike; flowers subtended by 1 membranous, lanceolate bract 3—5 mm long with a apical spine 1.5—3 mm long, and 2 smaller, similar bracteoles. **Calyx** of 5 unequal, lanceolate to elliptic sepals 2.5—4 mm long; the outer 2 the longest, keeled, with spines over 1.5 mm long, inner ones, acute-tipped. **Corolla** absent. Stamens 5, free. Ovary superior. **Fruit** a compressed, ovoid utricle 1.2—1.5 mm long, containing 1 shiny black seed.

Alternanthera pungens Kunth is native to tropical America and was first recorded from the Pacific Islands in 1959 (Hawai'i). It is locally common in disturbed, relatively dry lowland places, such as roadsides and lawns in Hawai'i (O'ahu, Moloka'i, Hawai'i), where it has spread rapidly since its relatively recent introduction, but is not recorded from the other Pacific islands.

This prostrate herb can be distinguished by its opposite, ovate to obovate leaves, sessile, ovoid, green to straw-colored, mostly glabrous spikes with spine-tipped bracts, and inconspicuous flowers. Synonyms: *Achyranthes repens* L., *Alternanthera repens* (L.) Link. A similar Hawaiian species, *Alternanthera caracasana* Kunth, differs in having densely pubescent bracts with spines less than 1 mm long.

ALTERNANTHERA SESSILIS
Amaranthaceae (Amaranth Family)

COMMON NAMES: sessile joyweed
DISTRIBUTION: all the main island groups

Prostrate to ascending, perennial herb. Stems up to 1 m long, rooting at the nodes, pubescent in longitudinal grooves and across the nodes. **Leaves** opposite, simple, blade elliptic to obovate or oblanceolate, 0.8—6 x 0.5—2 cm, acute at the apex, acuminate to cuneate at the base; surfaces glabrous; margins entire; petiole 1—4 mm long. **Inflorescence** of sessile, subglobose, axillary spikes 3—8 mm long. **Calyx** of 5 subequal, ovate to lanceolate sepals 1.5—3.5 mm long, white, glabrous, mostly 1-nerved, sessile, subtended by a lanceolate, membranous bract and 2 bracteoles up to *ca.* 1 mm long. **Corolla** absent. Stamens 5, 2 of which are sterile. Ovary superior, 1-celled. **Fruit** a compressed, ovoid or obcordate, pale brown utricle 2—2.3 mm long, 1-seeded.

Alternanthera sessilis (L.) R. Br. ex DC. is probably native to southern Asia, but is now pantropic in distribution. It was first recorded from the Pacific Islands in 1904 (Samoa), but is now found throughout the area. It is common in disturbed, relatively wet places such as roadside ditches, wetland taro patches, and on the edge of marshes, and extends up to 1000 m elevation.

This prostrate to ascending herb can be distinguished by its stems that root at the lower nodes, opposite, elliptic to oblanceolate leaves, sessile, axillary, subglobose spikes, white bracts and sepals, absence of a corolla, and a single tiny seed in the dry fruit. Synonym: *Alternanthera nodiflora* R. Br.

AMARANTHUS SPINOSUS
Amaranthaceae (Amaranth Family)

COMMON NAMES: spiny amaranth; pakai, kukū (Hawai'i)
DISTRIBUTION: all the main island groups except Tonga and Tahiti

Annual herb up to 1.5 m in height. Stems erect, often red, sub-glabrous, with a pair of divergent axillary spines 4—18 mm long. **Leaves** alternate, simple, blade ovate to oblanceolate, mostly 1—8 x 0.6—3.5 cm, acute and mucronate at the apex, acute to acuminate at the base; surfaces glabrous; petioles 0.5—5 cm long. **Inflorescence** of axillary clusters of unisexual flowers in the lower axils, and in terminal or upper axillary, simple or branched spikes up to 20 cm long; bracts linear to lanceolate, 2—3.5 mm long. **Calyx** of 5 lanceolate to spathulate sepals 1.2—2.5 mm long, subtended by 2 bracteoles. **Corolla** absent. Stamens 5, free. Ovary superior, stigmas 2. **Fruit** an ovoid utricle 1.5—2.5 mm long, splitting open by a circumscissile cap, with 1 shiny black seed.

Amaranthus spinosus L. is probably native to tropical America, but is now pantropic in distribution. It was first recorded from the Pacific Islands in 1928 (Hawai'i) and is common in disturbed lowland places, such as roadsides and abandoned land, especially in Hawai'i.

This erect herb can be distinguished by its paired, axillary spines, inflorescences of axillary clusters and terminal panicles or spikes, and small ovate fruit whose top splits off to release the shiny black seed. A related but spineless species in Hawai'i, Tahiti, Fiji, and Micronesia (but not Guam), *Amaranthus dubius* Mart. ex Thell., has a splitting fruit with the cap margin wrinkled. A similar spineless species in Hawai'i and Fiji, *Amaranthus hybridus* L., has a smooth cap margin.

AMARANTHUS VIRIDIS
Amaranthaceae (Amaranth Family)

COMMON NAMES: slender amaranth; pakai (Hawai'i); tubua (Fiji)
DISTRIBUTION: all the main island groups

Erect to decumbent, monoecious, annual herb 15—150 cm in height. Stems nearly glabrous, striate, reddish. **Leaves** alternate, simple, blade ovate to rhombic, mostly 1.5—8 x 0.8—7 cm, acute to narrowly obtuse and slightly emarginate at the apex, truncate to broadly cuneate at the base; surfaces mostly glabrous; margins entire; petiole 0.5—8 cm long. **Inflorescence** of compact, cymose clusters in lower axils, and spike-like panicles terminal and upper axillary; bracts tiny. **Calyx** of 3 oblong to spathulate sepals 1—1.5 mm long, striped green in the middle, subtended by 2 tiny bracteoles. **Corolla** absent. Stamens 3. Ovary superior. **Fruit** a rugose, compressed-globose utricle 1.2—1.5 mm long, indehiscent or rupturing irregularly, containing a shiny black, lens-shaped seed.

Amaranthus viridis L. is probably native to somewhere in the Old World tropics, but was an ancient introduction to the Pacific Islands. It is occasional in disturbed lowland places such as croplands, roadsides, and villages. The leaves can be cooked and eaten like spinach.

This erect to low herb can be distinguished by its reddish spineless stems, acute, often slightly notched leaf tips, tiny greenish flowers in clusters in the lower axils and in loose panicles above, tiny green fruits that do not split open by a cap, and shiny black, lens-shaped seeds. Synonym: *Amaranthus gracilis* Desf. ex Poiret. A similar weedy species in Hawai'i, Samoa, Tonga, and Tahiti, *Amaranthus lividus* L., has generally shorter stems, smooth fruits, and leaves that are distinctly notched at the apex.

SCHINUS TEREBINTHIFOLIUS
Anacardiaceae (Mango Family)

COMMON NAMES: Christmas berry
DISTRIBUTION: Hawai'i (cultivated in many other island groups)

Small dioecious tree or shrub up to 7 m or more in height. Stems glabrous. **Leaves** alternate, odd-pinnately compound, rachis 3—12 cm long, narrowly winged, leaflets 5—9 per pinna, opposite, terminal blade largest, mostly elliptic, 4—9 x 1.5—3.5 cm, acute to subround at the apex, acute to oblique at the base, sessile; surfaces glabrous; margins entire to serrate. **Inflorescence** of many-flowered terminal and upper axillary panicles 2—8 cm long. **Calyx** of 5 triangular sepals 0.5—1 mm long. **Corolla** of 5 white, oblong to ovate petals 1.2—2.5 mm long. Stamens 10 in male trees, vestigial in female trees. Ovary superior, vestigial in male trees; stigmas 3. **Fruit** a red, globose drupe 4—6 mm in diameter, with one brown, sticky, compressed-globose, lumpy seed 3.5—4.5 mm in diameter.

Schinus terebinthifolius Raddi is native to Brazil and was first recorded from the Pacific Islands in 1911 (Hawai'i). It is common to locally abundant in disturbed scrub areas of Hawai'i at up to 900 m elevation, and is a serious weed that often forms dense thickets. It is sometimes planted as an ornamental, and has recently been reported to be an escape from cultivation in American Samoa. The leaves and berries are sometimes used in wreaths in Hawai'i, hence the name Christmas berry.

This small tree can be distinguished by its peppery-smelling foliage, alternate, odd-pinnately compound leaves with 5—9 leaflets, short, dense panicles of tiny white flowers, and small red globose fruits containing a single sticky seed.

CENTELLA ASIATICA
Apiaceae (Carrot Family)

COMMON NAMES: Asiatic pennywort; pohekula (Hawai'i); togotogo, togo, moa (Samoa); tono (Tonga); tohetupou (Tahiti); tododro (Fiji)
DISTRIBUTION: all the main island groups

Creeping perennial herb with erect leaves. Stems glabrous, rooting at the nodes. **Leaves** alternate, simple, often appearing clustered at the nodes, blade broadly cordate to reniform, 1—7 x 1.5—7 cm, rounded at the apex, cordate at the base, often with a broad sinus; surfaces glabrous, palmately 5—9-veined; margins dentate to crenate; petiole 2—30 cm long. **Inflorescence** of 1—4 short, subsessile, axillary umbels, subtended by 2 ovate bracts 1.5—3 mm long, atop a peduncle 0.3—4 cm long. **Calyx** of 5 linear sepals *ca.* 0.5 mm long, subsessile. **Corolla** of 5 purplish, suborbicular petals 0.5—1 mm long. Stamens 5, free. Ovary inferior; styles 2. **Fruit** a ribbed, compressed-globose schizocarp 2—3 mm long, splitting at maturity into two 1-seeded mericarps.

Centella asiatica (L.) Urb. is native to somewhere in Asia, but was an ancient introduction to the Pacific Islands as far east as Samoa, and a European introduction to Tahiti and Hawai'i. It is occasional to locally common in various disturbed habitats such as pastures, houseyards, and wetland margins, extending up to 1700 m elevation. The leaves are widely used in native remedies in Polynesia for treating various ailments (Whistler 1992b).

This prostrate herb can be distinguished by its creeping stems that root at the nodes, erect clusters of kidney-shaped leaves on short to long petioles, short umbels of tiny inconspicuous flowers, and small, flattened-globose fruits with longitudinal ribs. Synonym: *Hydrocotyle asiatica* L.

SCHEFFLERA ACTINOPHYLLA
Araliaceae (Ginseng Family)

COMMON NAMES: octopus tree
DISTRIBUTION: Hawai'i, Fiji, Tahiti, and cultivated elsewhere

Terrestrial or epiphytic tree up to 10 m or more in height. Stems glabrous, bark rough, gray. **Leaves** alternate, palmately compound, leaflets 5—15, leaflet blades elliptic to ovate, largest ones 10—30 x 4—11 cm, shortly acuminate to notched at the apex, rounded to acute at the base; surfaces glabrous, waxy; margins entire; stipules caudate, 5—10 cm long; petiole 10—50 cm long, sheathing at the base. **Inflorescence** of spreading subterminal branches 30—100 cm long bearing many 8—12-flowered heads. **Calyx** of 5 tiny lobes, sessile, subtended by 4 wing-like bracteoles. **Corolla** of 5 reddish, valvate, caducous petals. Stamens 5, free. Ovary inferior, many-celled, with styles fused to form a ring. **Fruit** a black drupe fused to form a subglobose multiple fruit 6—10 mm long.

Schefflera actinophylla (Endl.) Harms is native to Australia and New Guinea and was first recorded from the Pacific Islands in 1900 (Hawai'i). It is occasional in disturbed, relatively wet lowland places such as shrubland and pastures. Although introduced to many islands as an ornamental, for which it is still commonly used, in the islands it is reported to be naturalized only in Hawai'i and Fiji.

This tree, which is sometimes epiphytic, can be distinguished by its large, palmately compound leaves, sheathing petiole bases, red flowering branches somewhat resembling the tentacles of an octopus, sessile flowers in heads, and black multiple fruits. Synonym: *Brassaia actinophylla* Endl.

AGERATUM CONYZOIDES
Asteraceae (Sunflower Family)

COMMON NAMES: ageratum; maile hohono (Hawai'i)
DISTRIBUTION: all the main island groups

Erect, sparingly branched annual herb mostly 20—70 cm in height. Stems pubescent, often purple; foliage fragrant. **Leaves** opposite, simple, blade ovate, 3—10 x 3—7 cm, acute at apex, shortly cuneate to truncate at base; surfaces pubescent, lower surface gland-dotted; margins crenate; petiole 1—5 cm long. **Inflorescence** a terminal corymb 1—6 cm long, ultimately in discoid heads 4—6 mm long surrounded by a campanulate involucre of 2 series of lanceolate bracts 3—4 mm long. **Ray florets** absent. **Disc florets** tubular, 1.7—2.4 mm long, white, tipped with lavender. **Fruit** a black, cylindrical, sparsely puberulent, 5-angled achene 1.5—2 mm long, with a terminal pappus of 5 acuminate scales slightly longer than the achene.

Ageratum conyzoides L. is native to tropical America, but is now widespread throughout the tropics. It was first recorded from the Pacific Islands in 1871 (Hawai'i), where it is common as a weed of disturbed places in wet and dry habitats from near sea level to 1300 m elevation.

This erect herb can be distinguished by its fragrant foliage, opposite, pubescent leaves, crenate leaf margins, corymbs of small heads containing lavender-tipped disc florets, and small black achenes bearing an equally long pappus of sharp-tipped scales. It is similar to *Ageratum houstonianum* Mill. of Hawai'i, Tahiti, and Fiji, which differs in its denser inflorescences, lack of glands on the lower leaf surface, and cordate to truncate leaf bases.

BIDENS ALBA
Asteraceae (Sunflower Family)

COMMON NAMES: beggar's tick
DISTRIBUTION: Hawai'i, Samoa, Guam, Belau

Erect annual herb mostly 20—80 cm in height. Stems 4-angled, glabrous. **Leaves** opposite, simple or more commonly deeply 3-lobed, 4—18 cm long, lobe or leaf blade ovate to lanceolate, terminal one mostly 3—10 x 0.8—4 cm, acute at the apex, decurrent at the base; surfaces mostly glabrous; margins serrate; petiole 1—5 cm long. **Inflorescence** of terminal and axillary cymes 5—22 cm long, ultimately in discoid heads 6—11 mm long surrounded by 2 series of lanceolate involucral bracts 5—6 mm long. **Ray florets** 5—8, white, 9—13 mm long, sterile. **Disc florets** many, corolla tubular, *ca.* 3.5—4.5 mm long, yellow; palea linear, 3—4 mm long. **Fruit** a black, longitudinally ribbed, straight or curved, linear achene mostly 6—8 mm long, with 2 barbed, terminal awns *ca.* 2 mm long.

Bidens alba (L.) DC. is native to tropical America, but is now widespread in the tropics and subtropics. It was first recorded from the Pacific Islands in 1958 (Hawai'i) and is common as a weed of disturbed places such as roadsides, lawns, and abandoned plantations up to 500 m elevation. It spreads by means of barbed achenes that adhere to clothing, feathers, or fur.

This erect herb can be distinguished by its opposite, simple or trifoliate leaves, heads of yellow disk florets, white ray florets 9—15 mm long, and cylindrical achenes with 2 barbed, terminal awns. The similar, widespread *Bidens pilosa* L. differs in having rays absent and fewer, or smaller (4—7 rays that are 2—8 mm long). *Bidens cynapiifolia* Kunth of Hawai'i differs in having 3 or 4 short yellow rays and 4-awned achenes.

CALYPTOCARPUS VIALIS
Asteraceae (Sunflower Family)

COMMON NAMES: calyptocarpus
DISTRIBUTION: Hawai'i, Guam

Prostrate or low herb. Stems mostly 2—6 cm long, pubescent, rooting at the nodes. **Leaves** opposite, simple, blade ovate, mostly 1—4 x 0.8—2 cm, acute at the apex, acute to rounded at the base; surfaces appressed-pubescent; margins serrate; petiole 2—15 mm long. **Inflorescence** of axillary or terminal, solitary heads surrounded by 3—5 ovate to oblong bracts 5—10 mm long, subsessile or on a peduncle up to 2 cm long. **Ray florets** 3—5, with yellow rays 1.5—3 mm long. **Disc florets** 3—5, with a yellow corolla 2—3 mm long; palea linear, as long as achene. **Fruit** an achene 3—4 mm long, flattened-obovoid, somewhat dimorphic, with a pair of spreading awns 1—2 mm long.

Calyptocarpus vialis Less. is native to tropical and subtropical America and was first recorded from the Pacific Islands in 1963 (Hawai'i). It is common in lowland disturbed places on all the main islands of Hawai'i, often dominating in lawns and other places where the vegetation is kept low. It has only recently been noted from Guam.

This prostrate herb can be distinguished by its small, opposite, ovate, appressed-pubescent leaves, solitary heads of small yellow ray and disc florets, and flattened, 2-awned achenes. Synonym: *Synedrella vialis* (Less.) A. Gray. It is very similar to *Synedrella nodiflora* (see p. 42), but can usually be distinguished by its smaller size and prostrate habit.

CHROMOLAENA ODORATA
Asteraceae (Sunflower Family)

COMMON NAMES: none
DISTRIBUTION: Guam, Belau

Scandent shrub. Stems up to 3 m or more in length, densely pubescent, striate, forming opposite spreading branches. **Leaves** opposite, simple, blade ovate, 4—14 x 2.5—10 cm, acuminate at the apex, broadly acute to truncate at the base; surfaces appressed-pubescent, 3-veined, lower side covered with tiny yellow glands; margins coarsely toothed; petiole 1—3 cm long. **Inflorescence** a terminal corymb of cylindrical heads that are 8—12 mm long and surrounded by 3—5 series of striate, ovate to narrowly oblong bracts. **Ray florets** absent. **Disc florets** up to 30 or more, corolla 4—5 mm long, lavender to white, style long-exserted. **Fruit** a black linear achene 3—4 mm long, scabrous on the angles, with a pappus of white hairs 4.5—5.5 mm long.

Chromolaena odorata (L.) King & Robinson is native to tropical America and was first recorded from the Pacific Islands prior to 1963 (Guam). It is common in lowland scrub forests, thickets, pastures, and other disturbed places in Guam and on some of the other Micronesian islands, but is absent from Polynesia and Fiji.

This sparingly branched shrub can be distinguished by its lateral branches forming in opposite pairs, opposite, strong-smelling leaves with toothed margins, terminal corymbs of heads, 3—5 series of bracts around the heads, and lavender to white disc florets with exserted styles. Synonym: *Eupatorium odoratum* L.

CONYZA BONARIENSIS
Asteraceae (Sunflower Family)

COMMON NAMES: hairy horseweed
DISTRIBUTION: all the main island groups (but rare in Samoa)

Erect perennial herb mostly 50—150 cm in height. Stems densely pubescent. **Leaves** alternate, simple, blade narrowly oblanceolate, basal ones up to 15 x 2 cm (upper ones smaller), acute at the apex, attenuate at the base, sessile; surfaces gray, woolly; margins coarsely serrate. **Inflorescence** of terminal and upper-axillary clusters of heads surrounded by an involucre of narrowly lanceolate bracts 3—6 mm long. **Ray florets** 50—200 per head, pistillate. **Disc florets** fewer, perfect, corolla of both tubular, 4—5 mm long, cream colored. **Fruit** a slightly laterally compressed cylindrical achene 1—1.5 mm long, with a pappus of *ca.* 15—20 slender, yellowish brown to reddish bristles about as long as the corolla.

Conyza bonariensis (L.) Cronq. is native to South America and was first recorded from the Pacific Islands in 1871 (Hawai'i). It is a common weed in dry to mesic disturbed habitats, such as roadsides and waste places, from near sea level up to 3000 m elevation.

This sparingly branched herb can be distinguished by its narrow, alternate, densely pubescent, coarsely toothed leaves, heads in terminal and upper axillary clusters, and achenes bearing a pappus of straight, slender bristles. Synonyms: *Conyza albida* Willd. ex Spreng., *C. floribunda* H.B.K., *Erigeron albidus* (Willd.) A. Gray, *E. bonariense* L. The similar but less hairy, widespread *Conyza canadensis* (L.) Cronq. has a shorter involucre (3—4 mm long) with fewer florets (25—40 per head).

CRASSOCEPHALUM CREPIDIOIDES
Asteraceae (Sunflower Family)

COMMON NAMES: none
DISTRIBUTION: all the main island groups

Erect annual herb mostly 30—50 cm in height. Stems longitudinally ribbed, glabrous or pubescent. **Leaves** alternate, simple, blade elliptic to ovate, 3—18 x 1—6 cm, acute at the apex, decurrent at the base; surfaces glabrous; margins serrate or irregularly lobed; petiole 0.6—3 cm long. **Inflorescence** of terminal and upper-axillary corymbs 4—30 cm long, bearing drooping cylindrical heads 10—16 mm long; involucre of numerous linear bracts 8—10 mm long, subtended by a series of tiny basal bracts. **Ray florets** absent. **Disc florets** tubular, many, 8—10 mm long, reddish brown at the tips. **Fruit** a dark brown, puberulent, 8—10-ribbed, cylindrical achene 1.8—2.2 mm long, with a pappus of many soft, white bristles *ca.* 12 mm long.

Crassocephalum crepidioides (Benth.) S. Moore is native to tropical Africa, but is now widespread in the Old World tropics. It was first recorded from the Pacific Islands in 1929 (Hawai'i), but has rapidly spread throughout the area. It is occasional to common in disturbed places, such as plantations, waste land, and even in open native forest, from the lowlands to 1750 m elevation.

This tall, erect herb can be distinguished by its irregularly serrate to pinnately lobed, alternate leaves, terminal corymbs of drooping heads bearing one series of linear bracts subtended by a series of tiny bracts at the base, reddish brown disk florets, and plumed achenes. A similar widespread species, *Erechtites valerianifolia* (Wolf) DC. differs in having more deeply lobed leaves, narrower, more numerous heads, and pink corollas.

ELEPHANTOPUS MOLLIS
Asteraceae (Sunflower Family)

COMMON NAMES: tobacco weed, elephant's foot
DISTRIBUTION: all the main island groups except Samoa

Coarse erect herb up to 150 cm in height. Stems pubescent, longitudinally ridged. **Leaves** alternate, simple, blade elliptic to obovate, 8—20 x 2—7 cm, acute at the apex, long-decurrent at the base; upper surface pubescent, lower densely so; margins crenate or serrate. **Inflorescence** an open panicle of subglobose clusters of heads 1—2 cm wide subtended by three triangular, leaf-like bracts, and each head 6—8 mm long surrounded by 8 brown, lanceolate bracts in two series. **Ray florets** absent. **Disc florets** tubular, white to pink, 3—4 mm long, usually 4 per head. **Fruit** a brown, glabrous, distinctly ribbed, linear-cylindrical achene 2—3 mm long, with a pappus of 5 straight white bristles almost as long as the achene.

Elephantopus mollis Kunth is native to tropical America, but is now pantropical in distribution. It was first recorded from the Pacific Islands in 1912 (Tahiti), where it is occasional to common in sunny, dry, disturbed places such as plantations and pastures. It spreads by means of its achenes that stick to clothing and fur, and is reported to be a serious weed in some countries.

This tall, erect herb can be distinguished by its alternate, hairy leaves, much-branched panicles, subglobose clusters of brown, narrowly cylindrical heads, and achene with 5 straight white bristles.

EMILIA FOSBERGII
Asteraceae (Sunflower Family)

COMMON NAMES: Flora's paintbrush
DISTRIBUTION: Hawai'i, Tahiti, Fiji, Guam

Annual herb up to over 60 cm in height. Stems pubescent to glabrous. **Leaves** alternate, simple, blade extremely variable (elliptic to oblong-lanceolate), mostly 4—15 x 1—7 cm, acute at the apex, lobed to winged and attenuate at the base; surfaces glabrous; margins coarsely dentate; petiole 0—5 cm long. **Inflorescence** a loose corymb of 1 to few urn-shaped heads; involucre of 10—13 linear-lanceolate bracts 9—14 mm long, reflexed when fruits are mature. **Ray florets** absent. **Disc florets** over 50, with red, narrowly tubular corollas 6—8 mm long exceeding the involucre by 2—5 mm. **Fruit** a cylindrical, 5-ribbed achene 2.5—3.5 mm long, brown.

Emilia fosbergii Nicolson is a probably native to Africa, where it arose from a hybridization between *Emilia sonchifolia* and *E. coccinea*, but is now pantropic in distribution. It was first recorded from the Pacific Islands in 1920 (Hawai'i), where it is occasional in disturbed places at up to 800 m elevation.

This weak-stemmed herb can be distinguished by its toothed, variable leaves, loose corymbs of urn-shaped heads, and red corollas that exceed the 10—13 narrowly lanceolate bracts. Synonym: *Emilia javanica* as incorrectly used by some authors. The widespread and common *Emilia sonchifolia* (L.) DC. differs in having lobed leaves, narrower heads, and purple or lavender corollas that scarcely exceed the bracts in length.

MIKANIA MICRANTHA
Asteraceae (Sunflower Family)

COMMON NAMES: mile-a-minute vine; fue saina (Samoa); wa bosucu (Fiji)
DISTRIBUTION: Samoa, Tonga (rare), Tahiti, Fiji

Scrambling or climbing, perennial herbaceous vine. Stems glabrous, ribbed. **Leaves** opposite, simple, blade cordate or deltoid, 4—10 x 2—5 cm, acute or acuminate at the apex, broadly cordate at the base; surfaces glabrous; margins crenate to wavy or subentire; petiole 2—5 cm long. **Inflorescence** of dense axillary or terminal corymbs, bearing discoid heads surrounded by an involucre of 4 subequal, dorsally ribbed bracts 3—4 mm long, with a smaller basal bract. **Ray florets** absent. **Disc florets** mostly 4 per head, corolla tubular, 5-lobed, 1.5—2.5 mm long, white. **Fruit** a black, 5-angled, linear-oblong achene 1.5—2 mm long, with a pappus of many white bristles 2.5—3.5 mm long.

Mikania micrantha Kunth is native to tropical America and was first recorded from the Pacific islands in 1906 (Fiji). It is abundant in most relatively moist, disturbed habitats up to 1750 m elevation. It is very common in Fiji and is the most prevalent weed in Samoa; in these two archipelagoes the leaves are commonly rubbed onto cuts and wounds to staunch the bleeding (Whistler 1992b).

This herbaceous vine can be distinguished by its opposite, triangular leaves, corymbs of small white, disc-shaped heads, and achenes with a pappus of white bristles. *Mikania scandens* (L.) Willd., which is virtually indistinguishable from *M. micrantha*, is common in Guam. If the two are actually the same species, the former name is the correct one.

PLUCHEA CAROLINENSIS
Asteraceae (Sunflower Family)

COMMON NAMES: sourbush, pluchea
DISTRIBUTION: Hawai'i, Society Islands, Guam, Belau

Erect shrub up to 4 m in height, but usually much less. Stems glandular-tomentose. **Leaves** alternate, simple, blade lanceolate to elliptic, 6—15 x 1—6 cm, acute at the apex and base; upper surface gray-green, glandular, sparsely pubescent, lower surface densely glandular-tomentose, pale; margins mostly entire; petiole 1—2.5 cm long. **Inflorescence** a terminal, broad, flat-topped panicle bearing numerous campanulate to cup-shaped heads; involucre 4.5—6 mm long, bracts in several series, outer ones ovate, inner ones lanceolate. **Ray florets** absent. **Disc florets** many, with a pink, filiform corolla 2—3 mm long. **Fruit** a cylindrical, appressed-pubescent achene less than 1 mm long, with a conspicuous pappus of *ca.* 12 white bristles 2—3 mm long.

Pluchea carolinensis (Jacq.) G. Don is native to tropical America, and was first recorded from the Pacific Islands in 1931 (Hawai'i). It is locally common in dry, often saline areas near the coast, but on occasion up to 900 m elevation. It is particularly common in Hawai'i, but is virtually absent from the South Pacific (except in the Society Islands).

This shrub can be distinguished by its alternate, lanceolate to elliptic leaves with a fuzzy lower surface, mostly entire leaf margins, flat-topped panicles of pink disk florets, and tiny achenes with long white pappus bristles. Synonyms: *Pluchea odorata* and *Pluchea symphytifolia* of some authors . It differs from *Pluchea indica*, which has shorter, mostly glabrous leaves with toothed margins.

PLUCHEA INDICA
Asteraceae (Sunflower Family)

COMMON NAMES: Indian pluchea, Indian fleabane
DISTRIBUTION: Hawai'i, Samoa (rare), Guam, Belau

Erect shrub up to 1.5 m or more in height. Stems sparsely pubescent when young, glabrous at maturity. **Leaves** alternate, simple, blade obovate, 2.5—4 x 1.2—2 cm, broadly acute to rounded at the apex, cuneate at the base, subsessile; surfaces glandular; margins coarsely toothed. **Inflorescence** a terminal, hemispherical panicle of cylindrical to campanulate heads; involucre 2.5—5 mm, with bracts in several imbricate series, outer ones ovate, inner ones lanceolate. **Ray florets** absent. **Disc florets** many, with a pink, filiform corolla 2.5—3.5 mm long. **Fruit** a cylindrical glabrous achene up to 1 mm long, with a conspicuous terminal pappus of *ca.* 16 white bristles 3—4 mm long.

Pluchea indica (L.) Less. is native to Southeast Asia, but is now widespread in the tropics. It was first recorded from the Pacific Islands in 1915 (Hawai'i), where it is locally common to abundant in lowland saline areas, particularly along estuaries and especially in Hawai'i, mostly near sea level, but occasionally up to 450 m elevation.

This shrub can be distinguished by its alternate, obovate leaves with toothed margins, hemispherical panicles of pink disc florets, and tiny achenes with long white pappus bristles. It readily hybridizes with *Pluchea carolinensis* to form an intermediate plant known as *Pluchea x fosbergii* Cooperr. & Galang where the two species occur together; this hydrid has been recorded from Hawai'i and Guam.

PSEUDELEPHANTOPUS SPICATUS
Asteraceae (Sunflower Family)

COMMON NAMES: false elephant's-foot; vao 'elefani (Samoa)
DISTRIBUTION: all the main island groups except Belau

Erect, tough-stemmed, deep-rooted annual herb 15—100 cm in height. Stems pubescent, ribbed. **Leaves** alternate, simple, blade oblanceolate to lanceolate, lower ones up to 3 x 12 cm, reduced in size on the upper stems, subsessile, acute at the apex, decurrent at the base; surfaces pubescent, lower surface gland-dotted; margins wavy. **Inflorescence** of terminal and axillary spikes of narrowly ellipsoid, discoid heads 8—12 mm long, subtended by a reduced leaf and surrounded by 4 pairs of bracts. **Ray florets** absent. **Disc florets** 4 per head, corolla 5.5—7 mm long, white to violet. **Fruit** a densely pubescent, narrowly obconical, 10-ribbed achene 5—7 mm long; pappus of 6—10 white, unequal bristles 4—8 mm long, the longest 2 with an S-shaped curve.

Pseudelephantopus spicatus (Juss. ex Aubl.) C.F. Baker is native to tropical America and was first recorded from the Pacific Islands in 1945 (Fiji). It is common as a weed of disturbed places, particularly in lawns and plantations, from near sea level up to 900 m elevation. Its wiry stems make it difficult to eradicate, and it is a declared noxious weed in Fiji.

This tough-stemmed herb can be distinguished by its alternate oblanceolate to lanceolate leaves, inflorescences of disc-shaped heads on narrow spikes, inconspicuous white disc flowers surrounded by four pairs of overlapping bracts, and achenes with bristles unequal and some S-shaped. Synonym: *Elephantopus spicatus* Juss. ex Aubl., which is the name used in the flora of Hawai'i (Wagner *et al.* 1990).

SONCHUS OLERACEUS
Asteraceae (Sunflower Family)

COMMON NAMES: sow thistle; pualele (Hawai'i); longolongo'uha (Tonga)
DISTRIBUTION: all the main island groups except Samoa and Belau

Erect annual herb up to 1.2 m or more in height, with a taproot. Stems glabrous or sparingly glandular-pubescent, ribbed; sap milky. **Leaves** alternate, simple, blade extremely variable, mostly pinnately lobed, blade 4—25 x 1—10 cm, acute at the tip, sessile, clasping, and forming a pair of auricles at the base; surfaces glabrous; margins dentate and spiny. **Inflorescence** a cyme of 1 to several ovoid heads; involucre 6—10 mm long, in two series of lanceolate bracts. **Ray florets** many, ray of corolla yellow, 5—7 mm long, 5-toothed at the tip. **Disc florets** absent. **Fruit** a brown, flattened-cylindrical achene 2—3 mm long, finely rugose, 6—10 ribbed, with a terminal pappus of many white bristles 5—9 mm long, some persistent, some caducous.

Sonchus oleraceus L. is native to Europe, but is now found throughout the tropics and subtropics. It could have been an ancient introduction to the Pacific Islands, since it was recorded from Tonga in 1773, but it arrived later in other islands (Hawai'i by 1871). It is occasional in disturbed habitats, particularly in croplands and lawns, at up to 900 m elevation.

This erect herb can be distinguished by its milky sap, sessile, clasping leaves, dentate, spiny, lobed leaf margins, cymes of heads with yellow ray florets, and small achenes with a terminal pappus of white bristles. The light, parachute-like achenes are readily dispersed by the wind. These are similar to those of several other Asteraceae species, but of these only *Youngia japonica* also has milky sap.

SYNEDRELLA NODIFLORA
Asteraceae (Sunflower Family)

COMMON NAMES: nodeweed, synedrella; tae'oti (Samoa); pakaka (Tonga)
DISTRIBUTION: all the main island groups

Coarse, erect herb mostly 10—70 cm in height. Stems appressed-hairy, ribbed. **Leaves** opposite, simple, blade ovate to elliptic, 1.5—8 x 0.8—5 cm, subsessile, acute at the tip, attenuate at the base; surfaces with hairs forming white dots at their base; margins subentire to crenate. **Inflorescence** of axillary heads 6—9 mm long, surrounded by an involucre of 2 series of lanceolate bracts 6—10 mm long; peduncle 1—10 mm long. **Ray florets** pistillate, mostly 3—5, ray yellow, 3-lobed, 1—2 mm long. **Disc florets** mostly 8 or 9, corolla tubular, 2—3 mm long, 4-lobed, yellow. **Fruit** a dark achene 4—5 mm long, those of the ray florets flat, oval, and spiny-edged with 2 terminal awns *ca.* 1 mm long, those of the disk florets cylindrical with 2 terminal awns *ca.* 2.5—5 mm long.

Synedrella nodiflora (L.) Gaertn. is native to tropical America, but is now pantropical in distribution. It was first recorded from the Pacific Islands in 1905 (Samoa), where it is common in disturbed habitats such as lawns, roadsides, and plantations up to 450 m elevation.

This coarse herb can be distinguished by its opposite leaves, axillary, solitary, subsessile heads bearing several yellow ray and disk florets, and two types of awned achenes. It is very similar to *Calyptocarpus vialis* (see p.31), but is more robust and erect. The similar *Eleutheranthera ruderalis* (Sw.) Sch.-Bip. of Samoa and Fiji differs in having smaller leaves and short-stalked, drooping heads.

TRIDAX PROCUMBENS
Asteraceae (Sunflower Family)

COMMON NAMES: coat buttons
DISTRIBUTION: all the main island groups

Ascending to procumbent perennial herb. Stems 10—50 cm long, hirsute, ribbed. **Leaves** opposite, simple, blade ovate to lanceolate, mostly 2—5 x 0.8—2.5 cm, acute at the apex and base; margins irregularly toothed or lobed; surfaces strigose; petiole 4—12 mm long, hairy. **Inflorescence** a terminal or axillary, solitary, campanulate head on a peduncle 8—20 cm long, surrounded by 2 series of *ca.* 10 involucral bracts, 3.5—6.5 mm long. **Ray florets** 3—6, pistillate, corolla strap-shaped, ray 2.5—4 mm long, 3-lobed, white or cream. **Disc florets** many, perfect, corolla tubular, 5—6.5 mm long, 5-lobed, yellow; palea membranous, as long as and enclosing the floret. **Fruit** a pubescent, black, turbinate achene 2—2.5 mm long, with a pappus of 20 unequal plumose bristles up to 5—7 mm long.

Tridax procumbens L. is native to tropical America and was first recorded from the Pacific Islands in 1906 (Fiji). It is common as a weed of disturbed, relatively dry places, particularly in waste areas and along roadsides in coastal areas and lowlands.

This low-growing herb can be distinguished by its opposite hairy leaves, toothed or lobed leaf margins, solitary, bell-shaped heads on ascending stalks 8—20 cm long, white, 3-lobed ray florets, yellow disc florets, and black, top-shaped achenes bearing plumose bristles.

VERBESINA ENCELIOIDES
Asteraceae (Sunflower Family)

COMMON NAMES: golden crown-beard
DISTRIBUTION: Hawai'i

Much-branched annual herb up to 1 m in height. **Leaves** opposite near the base, upper ones alternate, simple, blade lanceolate to deltoid, 4—12 x 2—10 cm, acute at the apex, acute to truncate at the base; upper surface green, pubescent; lower surface gray, canescent; margins coarsely toothed; petiole with a pair of basal auricles. **Inflorescence** of a solitary or several subglobose heads on long peduncles; involucre of *ca.* 15 lanceolate bracts 7—15 mm long. **Ray florets** 10—15, with an oblanceolate, 3-toothed, bright yellow corolla 1—2.5 cm long. **Disc florets** numerous, with a yellow tubular corolla 2—3 mm long, with linear chaffy bracts. **Fruit** a flattened, narrowly winged, deltoid achene *ca.* 3 mm long, with 2 terminal pappus awns.

Verbesina encelioides (Cav.) Benth. & Hooker is native to Mexico and the southeastern U.S., and was first recorded from the Pacific Islands in 1871 (Hawai'i). It is common in disturbed, dry lowland areas, but occasionally inland at up to 2800 m elevation in Hawai'i.

This large, widely branching herb can be distinguished by its opposite or alternate, deltoid to lanceolate leaves, gray lower leaf surfaces, petioles with a pair of ear-like appendages at the base, subglobose heads on long stalks, and yellow disc florets and large yellow ray florets that make it look very similar to a sunflower.

VERNONIA CINEREA
Asteraceae (Sunflower Family)

COMMON NAMES: little ironweed
DISTRIBUTION: all the main island groups

Erect, loosely branching annual herb mostly 15—60 cm in height. Stems finely pubescent, ribbed. **Leaves** alternate, simple, blade oblanceolate to obovate, 2—5 x 0.8—2 cm, reduced on upper stems, subsessile, acute to obtuse at the apex, decurrent at the base; surfaces gland-dotted; margins irregularly toothed. **Inflorescence** of loose corymbose cymes 2—13 cm long, bearing many campanulate heads surrounded by several series of lanceolate involucral bracts, the inner ones 3—4 mm long, reflexed at maturity. **Ray florets** absent. **Disc florets** 20—25, corolla tubular, 3—4 mm long, 5-lobed, lavender. **Fruit** a cylindrical, pubescent achene 1.5—2 mm long, with a pappus of numerous, persistent, spreading white bristles 3—4 mm long.

Vernonia cinerea (L.) Less. is native to tropical America, but is now widespread throughout the tropics. It was first recorded from the Pacific Islands in 1871 (Hawai'i), where it is occasional in relatively dry, disturbed places of both low and high islands, extending up to 850 m elevation.

This erect herb can be distinguished by its loosely branching habit, small, alternate, ovate leaves, loose cymes of small heads, lavender disc florets, and a pappus of numerous white bristles on the achene. Synonyms: *Conyza cinerea* L., *Vernonia parviflora* Reinw.

WEDELIA TRILOBATA
Asteraceae (Sunflower Family)

COMMON NAMES: wedelia
DISTRIBUTION: Hawai'i, Samoa, Tahiti, Guam, Belau

Creeping perennial herb. Stems up to 40 cm long, ascending when flowering, coarsely hairy when young, rooting at the nodes. **Leaves** opposite, simple, blade somewhat fleshy, obovate to elliptic or ovate, 2—9 x 1.2—4 cm, acute at the apex, rounded to cuneate, winged and sessile at the base; surfaces appressed-pubescent; margins irregularly toothed. **Inflorescence** of terminal and axillary, solitary heads on a peduncle 2—9 cm long, involucre campanulate, with 2—4 series of bracts. **Ray florets** mostly 8—13, 6—15 mm long, 2- or 3-toothed, yellow, pistillate. **Disc florets** numerous, tubular, yellow, 4—5 mm long, mixed with chaffy bracts. **Fruit** a 2—4 angled, tuberculate achene 4—5 mm long, with short, narrow pappus scales on the top.

Wedelia trilobata (L.) Hitchc. is native to tropical America and was first recorded from the Pacific Islands sometime before 1965 (Hawai'i). It is commonly grown as an ornamental ground cover, but is naturalized in disturbed places in Hawai'i, at least on O'ahu and Kaua'i, and to a lesser extent in American Samoa and Micronesia, usually in relatively wet places in the lowlands. It often dominates where it occurs, and spreads vegetatively since it does not produce seeds.

This prostrate herb can be distinguished by its creeping stems, opposite, sessile, somewhat fleshy leaves, toothed leaf margins, and showy yellow composite inflorescences of yellow ray and disc florets.

XANTHIUM STRUMARIUM
Asteraceae (Sunflower Family)

COMMON NAMES: cocklebur; kīkānia (Hawai'i)
DISTRIBUTION: Hawai'i, Tahiti

Coarse, erect annual subshrub up to 2 m in height. Stems pubescent with short, thick hairs. **Leaves** alternate, simple, blade broadly ovate to reniform, often 3-lobed, 3—13 x 2.5—12 cm, acute to obtuse at the apex, mostly cordate at the base; surfaces scabrous, glandular; margins coarsely and irregularly toothed; petiole 2—15 cm long. **Inflorescence** in axillary, subsessile, pistillate, bur-like heads and terminal panicles of staminate heads 5—9 mm long. **Ray florets** absent. **Disc florets** of staminate heads with several tiny green florets subtended by 1—3 series of distinct bracts; disc florets of pistillate heads lacking corollas and pappus, subtended by the bur-like involucre. **Fruit** a brown cylindrical bur 1.5—3.5 cm long, with hooked prickles up to 7 mm long and 2 terminal beaks.

Xanthium strumarium L. is native to tropical America and was first recorded from the Pacific Islands in 1871 (Hawai'i). It is occasional to locally common in relatively dry, disturbed lowland places such as pastures and roadsides, and is a serious pest because of its large burs that stick to animal fur.

This coarse annual subshrub can be distinguished by its large, irregularly toothed leaves, inconspicuous male and female flowers in axillary heads, and relatively large bur-like fruits covered with hooked prickles. Synonym: *Xanthium saccharatum* Wallr. A related species in Fiji, *Xanthium pungens* Wallr., is also a serious pest.

YOUNGIA JAPONICA
Asteraceae (Sunflower Family)

COMMON NAMES: Oriental hawksbeard
DISTRIBUTION: all the main island groups (but not in Western Samoa) except Belau

Annual herb mostly 10—50 cm in height. Stems glabrous to sparsely pubescent, ribbed; sap milky. **Leaves** mostly basal, simple, blade oblanceolate to lyrate-pinnatifid, mostly 2—15 x 1—4 cm, subsessile, acute to rounded at the apex, narrowly winged at the base; margins subentire to pinnately lobed; surfaces glabrous. **Inflorescence** a terminal scapose cyme up to 50 cm or more in length, bearing many tiny, strap-shaped heads surrounded by an involucre of *ca.* 5 outer ovate bracts 0.5—1 mm long and *ca.* 8 linear-lanceolate bracts 3.5—5 mm long. **Ray florets** 10—20 per head, rays 3—5 mm long, yellow, 4—5-toothed. **Disc florets** absent. **Fruit** a brown, fusiform, ribbed achene 1.5—2.5 mm long, with a pappus of numerous white bristles 3—4 mm long.

Youngia japonica (L.) DC. is native to Southeast Asia and was first recorded from the Pacific Islands in 1864 (Hawai'i). It is occasional in moist, shaded, disturbed places, such as along sidewalks, roadsides, and in abandoned plantations, extending up to 1400 m elevation. It is readily spread by its parachute-like achenes.

This erect herb can be distinguished by its milky sap, rosette of basal, toothed to lyrate-pinnatifid leaves, delicate, widely branching, long-stalked cymes, two series of involucral bracts, yellow ray florets, and tiny achenes bearing numerous bristles. Synonym: *Crepis japonica* (L.) Benth.

SPATHODEA CAMPANULATA
Bignoniaceae (Jacaranda Family)

COMMON NAMES: African tulip-tree
DISTRIBUTION: all the main island groups (only cultivated in Tonga)

Large tree up to 25 m in height. Stems puberulent when young. **Leaves** opposite, odd-pinnately compound, rachis mostly 30—45 cm long, leaflets 9—19, blade elliptic, 7—13 x 3—6 cm, shortly acuminate at the apex, obtuse and oblique at the base; margins entire; stipules leaf-like, ovate to cordate, *ca.* 3 x 2 cm. **Inflorescence** a dense terminal raceme 8—25 cm long with the flowers crowded at the top. **Calyx** spathe-like, 4.5—6 cm long, splitting open along the side, densely brown tomentose, on a pedicel 1.5—4 cm long. **Corolla** campanulate, 9—13 cm long, slightly two-lipped, red-orange. Stamens 4, epipetalous. Ovary superior. **Fruit** an oblong capsule 16—24 x 3.5—6 cm, with the valves keeled. Seeds numerous, 2—2.5 cm long, including the membranous wings.

Spathodea campanulata P. Beauv. is native to central Africa and was first recorded from the Pacific Islands prior to the turn of the century (Hawai'i). It is commonly cultivated as an ornamental tree because of its showy orange-red flowers, but escapes and becomes naturalized in disturbed lowland areas, and such as pastures and shrublands in Fiji, Samoa, Hawai'i, and Guam and perhaps elsewhere as well.

This large large tree can be distinguished by its opposite, pinnately compound leaves, 9—13 large leaflets, spathe-like calyx covered with velvety, rusty-brown hairs, large, bell-shaped, showy orange-red flowers, and long pods that split open to release the winged seeds.

HELIOTROPIUM PROCUMBENS
Boraginaceae (Heliotrope Family)

COMMON NAMES: none
DISTRIBUTION: Hawai'i, American Samoa, Guam

Erect to decumbent perennial herb up to 50 cm in height, but usually much less. Stems branching, densely appressed-strigose. **Leaves** alternate, simple, blade narrowly oblanceolate to elliptic, 1—3 x 0.2—0.8 cm, acute or short-acuminate at the apex, attenuate at the base; surfaces densely appressed-strigose, gray to pale green; margins entire, sometimes revolute; petiole 3—12 mm long. **Inflorescence** in clusters of 1—3 loose scorpioid cymes 2—8 cm long. **Calyx** 1—2 mm long, deeply divided into 5 unequal, linear to lanceolate lobes, strigose. **Corolla** funnelform, white, 1.5—3 mm long, shallowly divided into 5 rounded lobes. Stamens 5, epipetalous, included. Ovary superior, 4-lobed. **Fruit** subglobose, 1—1.7 mm long, strigose, splitting into four 1-seeded nutlets.

Heliotropium procumbens Mill. is native to the southern U.S. and tropical America, and was first recorded from the Pacific Islands in 1905 (Guam). In Hawai'i and American Samoa it is uncommon to occasional in dry coastal areas, rarely extending very far inland, but is more common in Guam.

This prostrate herb can be distinguished by its gray to pale green, hairy alternate leaves, flowers in scorpioid cymes, tiny white, funnel-shaped corollas up to 3 mm long, and tiny round fruits that split into 4 nutlets. Synonym: *Heliotropium ovalifolium* var. *depressum* of Stone (1970) in Guam. A similar species in Hawai'i and Fiji, *Heliotropium amplexicaule* Vahl, has larger bluish flowers. See Whistler (1992a) for two similar native littoral species of the Pacific, *Heliotropium anomalum* and *H. curassavicum*.

OPUNTIA FICUS-INDICA
Cactaceae (Cactus Family)

COMMON NAME: prickly pear cactus; pānini (Hawai'i)
DISTRIBUTION: Hawai'i

Tree up to 5 m in height with a distinct trunk. Stems segmented, flattened, succulent, broadly obovate, 25—60 x 20—40 cm, with dull gray joints; spines absent or 1—6 at the areoles, white or yellowish, 1—3 cm long. **Leaves** subulate, small, caducous and not usually found on the plant. **Inflorescence** of solitary flowers borne at an areole, 6—7 cm long and 5—7 cm in diameter. **Calyx** and **corolla** of similar perianth parts, outer ones yellow with a red or green median stripe, 1—2 x 1.5—2, inner ones yellow to orange-yellow, 2.5—3 x 1.5—2 cm. Stamens numerous, filaments yellow. Ovary superior, with 8—10 stigma lobes. **Fruit** a white to reddish purple, ovoid berry 5—10 cm long containing numerous seeds.

Opuntia ficus-indica (L.) Mill. is probably native to Mexico, but is now widely cultivated and naturalized in the tropics and subtropics. It was first recorded from the Pacific Islands in about 1809 (Hawai'i), where it is locally common in relatively dry, disturbed lowland places on most of the the main islands of Hawai'i.

This cactus can be distinguished by its flat, succulent stems, absence of leaves, 1—6 spines per areole, numerous yellow petals, numerous stamens, and large white to purple, fleshy berries. Synonym: *Opuntia megacantha* Salm-Dyck. A similar species naturalized in Hawai'i and Micronesia, *Opuntia cochenillifera* (L.) Mill. (= *Nopalea cochenillifera* (L.) Salm-Dyck), differs in having red flowers and long-exserted styles and stamens.

ATRIPLEX SEMIBACCATA
Chenopodiaceae (Goosefoot Family)

COMMON NAMES: Australian saltbush
DISTRIBUTION: Hawai'i

Prostrate, perennial monoecious herb. Stems up to 1.5 m long, often red, forming spreading mats from the long taproot. **Leaves** alternate, simple, blade elliptic to ovate or spathulate, 0.5—2.5 x 0.3—1.2 cm, acute to rounded at the apex, acuminate to attenuate at the base; surfaces mealy; margins entire to toothed; petiole 1—3 mm long. **Inflorescence** of several tiny unisexual flowers in axillary clusters. **Calyx** tiny, shallowly divided into 3—5 lobes, surrounded in female flowers by 2 rhombic, strongly veined bracts. **Corolla** absent. Stamens 3—5 in male flowers. Ovary superior in female flowers. **Fruit** a 1-seeded utricle surrounded by 2 red, rhombic, fleshy bracts 4—6 mm long.

Atriplex semibaccata R. Br. is native to Australia and was first recorded from the Pacific Islands in 1895 (Hawai'i). It is common to locally abundant in dry or seasonally wet places in the lowlands of all the main islands of Hawai'i, reported up to 150 m elevation.

This prostrate herb can be distinguished by its reddish stems, small, narrow, alternate leaves with somewhat mealy surfaces, subentire to remotely toothed margins, axillary clusters of greenish flowers, and rhombic, fleshy red fruits. A similar Australian species occasional in Hawai'i, *Atriplex suberecta* Verd. (= *A. muelleri* of some Hawaiian authors) can be distinguished by its smaller yellowish green to pale brown fruits.

CHENOPODIUM MURALE
Chenopodiaceae (Goosefoot Family)

COMMON NAMES: goosefoot; ‘āheahea (Hawai’i)
DISTRIBUTION: Hawai’i, Tonga (rare)

Erect to ascending annual herb up to 1 m in height. Stems mealy pubescent, especially on young parts. **Leaves** alternate, simple, ovate to rhombic, 1.5—8 x 0.6—5 cm, acute at the apex, truncate to cuneate at the base; surfaces mealy-pubescent, at least on lower surface; margins irregularly and coarsely toothed; petiole 2—25 mm long. **Inflorescence** of small, dense clusters of flowers arranged in terminal and axillary, branched cymes up to 5 cm long. **Calyx** deeply divided into 5 oblong sepals 0.8—1.2 mm long, not completely enclosing the fruit. **Corolla** absent. Stamens 1—5, free. Ovary superior. **Fruit** a subglobose, 1-seeded utricle 1.2—1.5 mm long, containing a single black, lens-shaped seed with a very finely pitted surface.

Chenopodium murale L. is native to the Mediterranean region of Asia, but is now cosmopolitan in distribution. It was first recorded from the Pacific Islands in 1864 (Hawai’i), and it is occasional in disturbed places of Hawai’i, such as roadsides and the edges of coastal wetlands, but sometimes up to 2700 m elevation.

This erect herb can be distinguished by its alternate, rhombic to ovate leaves, mealy surfaces, dentate leaf margins, terminal and axillary cymes of inconspicuous gray-green flowers, and single black seed. A similar species from Hawai’i, Fiji, and Guam, *Chenopodium ambrosioides* L., has linear to spoon-shaped leaves and clusters of flowers in spikes or panicles. Another from Hawai’i, *Chenopodium carinatum* R. Br., is shorter and has smaller leaves.

IPOMOEA ALBA
Convolvulaceae (Morning-glory Family)

COMMON NAMES: moon flower; koali pehu (Hawai'i)
DISTRIBUTION: all the main island groups except Belau

Herbaceous to subwoody vine. Stems up to 6 m or more in length, glabrous, smooth or with soft prickles; sap milky. **Leaves** alternate, simple, blade broadly ovate, 8—16 x 8—14 cm, acute to attenuate at the apex, cordate at the base; surfaces glabrous; margins entire or sometimes 3—7-lobed; petiole 10—25 cm long. **Inflorescence** of 1—several flowers in axillary cymes up to 25 cm or more in length. **Calyx** of 5 fleshy, ovate to elliptic sepals 1—2 cm long, the inner 3 mucronate at the apex, the outer two caudate for up to half their length. **Corolla** salverform, white with green nectar guides, tube 7—12 cm long, limb 8—15 cm across. Stamens 5, epipetalous. Ovary superior. **Fruit** an ovoid to subglobose capsule 2—3 x 1—2 cm, dark brown, apiculate at the apex, 1—4 seeded.

Ipomoea alba L. is native to Mexico, but is now pantropic in distribution. It was first recorded from the Pacific Islands in 1819 (Hawai'i), probably originally being introduced as an ornamental vine, but is now naturalized. It is uncommon to locally common in disturbed places, especially on the margins of wetland areas in Hawai'i, ranging up to 600 m elevation.

This perennial vine can be distinguished by its soft prickles on the stem, milky sap, alternate, heart-shaped leaves, white corolla with a long tube and spreading limb, and brown, subglobose to ovoid capsule. The long-tubed white corolla serves to distinguish it from the other morning-glory species of the area except the native coastal vines *Ipomoea macrantha* and *Ipomoea tuboides*, which lack the soft prickles.

IPOMOEA CAIRICA
Convolvulaceae (Morning-glory Family)

COMMON NAMES: ivy-leaved morning glory; koali ʻai (Hawaiʻi)
DISTRIBUTION: Hawaiʻi, Tonga, Fiji

Perennial twining vine. Stems up to 4 m in length, mostly glabrous or with hairs at the nodes. **Leaves** alternate, simple, blade ovate to round, deeply divided into 5—7 lobes; lobes lanceolate to elliptic, 2—5 x 0.2—1.5 cm, acute to obtuse at the apex; surfaces glabrous; margins entire; petiole 1.5—6 cm long. **Inflorescence** an axillary, 1- to several-flowered cyme on a peduncle 0.5—8 cm long. **Calyx** 9—12 mm long, deeply divided into 5 ovate sepals with a short mucro at the tip, glabrous. **Corolla** funnel-shaped, 4.5—7 cm long, reddish purple, or rarely white with purple center. Stamens 5, epipetalous, included in the corolla throat. Ovary superior. **Fruit** a subglobose capsule 1—1.2 cm long, glabrous, 4-seeded.

Ipomoea cairica (L.) Sweet is native to tropical America, but is now pantropic in distribution. It was first recorded from the Pacific Islands in 1819 (Hawaiʻi), where it is now occasional in disturbed, mostly lowland places, such as roadsides and *Leucaena* scrub forest in Hawaiʻi and Fiji, at up to 670 m elevation. It is less common in Tonga, where it is more of an escaped ornamental. The roots can reportedly be used for food.

This perennial vine can be distinguished by its glabrous, alternate, deeply palmately lobed leaves, large, showy, purplish red, funnel-shaped corollas, and small subglobose capsules. Some other species of the family have similar "morning-glory" flowers, but these have heart-shaped leaves. Synonyms: *Ipomoea coptica* of some authors, *I. tuberculata* Roem. & Schult.

IPOMOEA INDICA
Convolvulaceae (Morning-glory Family)

COMMON NAMES: koali ʻawa (Hawai'i)
DISTRIBUTION: all the main island groups (but rare in Samoa)

Herbaceous to subwoody twining vine. Stems up to 10 m in length, pubescent, twining at the tips. **Leaves** alternate, simple, blade broadly ovate, 5—12 x 3—10 cm, acuminate to obtuse at the apex, cordate at the base; surfaces densely pubescent, especially on the lower side; margins entire; petiole 1.5—12 cm long, pubescent. **Inflorescence** a congested, few-flowered axillary cyme on a peduncle 1.5—15 cm long. **Calyx** 1.8—3.3 cm long, deeply cut into 5 lanceolate sepals acuminate to long-attentuate at the tip. **Corolla** funnel-shaped, 6—9 cm long, 6—8 cm in diameter, blue, purple, or pink (rarely white). Stamens 5, epipetalous. Ovary superior. **Fruit** a subglobose, usually angled, brown capsule up to 1 cm in diameter, 4-seeded.

Ipomoea indica (Burm.) Merr. is native to the Pacific Islands, and is pantropic in distribution. It is occasional to common in relatively dry, disturbed or open areas, sometimes in open native forest, mostly in the lowlands, but reported up to 1200 m elevation. It is particularly common in Hawai'i, where it has frequently been used in native medicines.

This perennial vine can be distinguished by its fuzzy foliage, heart-shaped leaves, long-stalked congested cymes, narrow bracts below the calyx, sepals with attenuate tips, and pink to blue corollas. Synonyms:*Ipomoea acuminata* (Vahl) Roem. & Schult., *I. congesta* R. Br., *Pharbitis insularis* Choisy. The widespread *Ipomoea littoralis* Bl. differs in its smaller leaves and flowers, glabrous foliage, and no bracts.

IPOMOEA OBSCURA
Convolvulaceae (Morning-glory Family)

COMMON NAMES: none
DISTRIBUTION: Hawai'i, Tahiti, Fiji, Guam

Perennial herbaceous vine. Stems up to 3 m in length, glabrous to pubescent, twining at the tip; sap clear. **Leaves** alternate, simple, blade ovate to cordate, 2.5—7 x 1.5—7 cm, acuminate or apiculate at the apex, cordate at the base; surfaces mostly glabrous; margins entire; petiole 1—8 cm long. **Inflorescence** axillary, solitary or in few-flowered cymes on a peduncle 1—4 cm long. **Calyx** 4—7 mm long, with 5 ovate sepals with the tip acute or apiculate; pedicel 1—2.5 cm long. **Corolla** campanulate, 1.5—2.5 cm long, 2.5—3 cm in diameter, white or cream-colored with yellow lines and a purple center. Ovary superior. **Fruit** a glabrous, globose capsule 8—12 mm long, with a persistent style, and containing 4 black seeds *ca.* 5 mm in diameter.

Ipomoea obscura (L.) Ker-Gawl. is native to somewhere in tropical Africa or Asia, but is now widespread in the Old World tropics. It was first recorded from the Pacific Islands in 1909 (Hawai'i), where it is occasional in dry, disturbed lowland places, such as roadsides and waste places, ranging up to 250 m elevation.

This herbaceous vine can be distinguished by its mostly glabrous stems and foliage, small, unlobed, alternate, heart-shaped leaves, and white, bell-shaped corollas dark in the center. It it is similar to *Ipomoea triloba* (see p. 58), which differs in having mostly lobed leaves and a purple corolla.

IPOMOEA TRILOBA
Convolvulaceae (Morning-glory Family)

COMMON NAMES: little bell
DISTRIBUTION: Hawai'i, Guam, Belau

Perennial herbaceous vine. Stems up to 4 m in length, glabrous or finely pubescent, twining at the tips; sap clear. **Leaves** alternate, simple, blade ovate to orbicular, sometimes palmately lobed, 2—10 x 2—8 cm, obtuse to acute at the apex, cordate at the base; surfaces mostly glabrous; margins entire or toothed; petiole 2—6 cm long. **Inflorescence** a compact, axillary, 1-to-few-flowered cyme on a peduncle 2—10 cm long. **Calyx** of 5 oblong to elliptic sepals 6—10 mm long, with ciliate margins and an obtuse to acute tip. **Corolla** funnel-shaped, purple (rarely white), 1.5—2.2 cm long. Stamens 5, epipetalous, included. Ovary superior. **Fruit** a dark brown, pubescent, subglobose capsule 5—6 mm long, containing 4 dark brown seeds 3—3.5 mm in diameter.

Ipomoea triloba L. is native to the West Indies, but is now widespread in the tropics. It was first recorded in the Pacific Islands in 1943 (Hawai'i), where it is occasional to common in disturbed lowland areas such as waste places and roadsides.

This herbaceous vine can be distinguished by its mostly glabrous stems and foliage, small, alternate, heart-shaped leaves usually with lobed or toothed margins, small flowers with a purple, bell-shaped corolla, and small capsule. It differs from other local purple-flowered species of the morning glory family in having smaller leaves and corollas.

MERREMIA AEGYPTIA
Convolvulaceae (Morning-glory Family)

COMMON NAMES: hairy merremia; koali kua hulu (Hawai'i)
DISTRIBUTION: Hawai'i, Guam

Perennial herbaceous vine. Stems up to 4 m or more in length, densely covered with white to yellowish hairs. **Leaves** alternate, palmately compound into 5 leaflets, blades elliptic, mostly 2—12 x 1—4 cm, acuminate to acute at the apex and base; surfaces long-hairy; margins mostly entire; petiole 3—14 cm long, long-hairy. **Inflorescence** an axillary, 1—3-flowered cyme on a long-hairy peduncle 2—15 cm long. **Calyx** of 5 ovate sepals 1—1.5 cm long, long-hairy. **Corolla** campanulate, 2—3 cm long, shallowly 5-lobed. Stamens 5, epipetalous, included in the corolla throat. Ovary superior. **Fruit** a brown, subglobose capsule 1—2 cm in diameter, surrounded by the enlarged calyx and containing 4 seeds.

Merremia aegyptia (L.) Urb. is pantropic in distribution, but is probably not native to the Pacific Islands, although it was first recorded there as early as the late 18th century (Hawai'i). It is common in disturbed places, including *Leucaena* scrub forest in the lowlands of Hawai'i, and is reported from all the main islands except Ni'ihau, at up to 350 m elevation.

This vine can be distinguished by its hairy stems, alternate, palmately compound leaves, white, bell-shaped corolla 2—3 cm long, and brown capsule 1—2 cm in diameter. Synonym: *Operculina aegyptia* (L.) House. *Ipomoea cairica (*see p. 55) has somewhat similar leaves, but the stems are glabrous and the corollas purple.

MERREMIA PELTATA
Convolvulaceae (Morning-glory Family)

COMMON NAMES: fue lautetele (Samoa); lagun (Guam)
DISTRIBUTION: all the main island groups except Hawai'i

Coarse climbing vine. Stems up to 20 m in length, nearly glabrous, twining at the tip; sap milky. **Leaves** alternate, simple, blade broadly ovate to round, 8—25 x 8—20 cm, acuminate at the tip, cordate or rounded at the base, peltate; surfaces nearly glabrous, veins of lower surface purple; margins entire; petiole 4—20 cm long. **Inflorescence** a loose, axillary, 5—15-flowered cyme 10—30 cm long. **Calyx** 1.8—3 cm long, deeply divided into 5 glabrous, ovate sepals; pedicel 1.5—5 cm long. **Corolla** funnel-shaped, shallowly 5-lobed, 4.5—7 cm long, white or rarely yellow. Stamens 5, epipetalous, included in the throat. Ovary superior. **Fruit** an ovoid capsule 1.5—2.5 cm long, enclosed within the dry, papery calyx, and containing 1—4 furry seeds.

Merremia peltata (L.) Merr. is native to the Pacific Islands and ranges from East Africa to Tahiti. It is common to abundant in disturbed places (especially in Samoa and Fiji), such as fallow land, but is also frequent in disturbed native forest up to 300 m elevation. It climbs over young trees and up into the canopy, slowing forest regeneration with its shade.

This high-climbing perennial vine can be distinguished by its milky sap, large, alternate, peltate leaves with an attenuate tip, large, white to yellow, funnel-shaped flowers, and papery sepals surrounding the fruit. Synonyms: *Ipomoea peltata* (L.) Choisy, *Merremia nymphaeifolia* (Dietr.) Hall. f. A related species in American Samoa and elsewhere, *Merremia umbellata* (L.) Hall. f., has smaller yellow flowers and non-peltate leaves.

MERREMIA TUBEROSA
Convolvulaceae (Morning-glory Family)

COMMON NAMES: wood rose; pilikai (Hawai'i)
DISTRIBUTION: Hawai'i, Fiji, Tahiti, and cultivated elsewhere

Liana climbing into trees. Stems up to 10 m or more long, glabrous. **Leaves** alternate, simple, blade suborbicular, deeply palmately 7-lobed, mostly 5—18 x 8—13 cm, lobes lanceolate to elliptic, acuminate at the apex; surfaces glabrous; margins of lobes entire; petiole 3—15 cm long. **Inflorescence** axillary, of solitary flowers or few-flowered compound cymes up to 20 cm long. **Calyx** of 5 unequal oblong sepals, inner ones 1.2—2 cm, outer ones 2.5—3 cm, subtended by 2 oblong bracts as long as outer sepals; pedicel 2—3 cm long. **Corolla** campanulate to nearly funnelform, shallowly 5-lobed, yellow, 5—6.5 cm long. Stamens 5, epipetalous. Ovary superior. **Fruit** a brown, subglobose, 1—4-seeded capsule 3—3.5 cm long, surrounded by the enlarged (up to 6 cm long) woody sepals.

Merremia tuberosa (L.) Rendle is native to tropical America and was first recorded from the Pacific Islands in 1932 (Hawai'i). It is occasional to locally common climbing in trees in relative moist places in Hawai'i at up to 550 m elevation, but is also cultivated there and in many of the other island groups. The enlarged woody sepals make this plant attractive in dry floral arrangements.

This high-climbing woody vine can be distinguished by its alternate, palmately lobed leaves, showy yellow, bell-shaped flowers, and enlarged woody sepals that surround the fruit. Synonym: *Ipomoea tuberosa* L.

OPERCULINA VENTRICOSA
Convolvulaceae (Morning-glory Family)

COMMON NAMES: pālulu (Samoa); fue hina? (Tonga); alalag (Guam)
DISTRIBUTION: Samoa, Tonga, Guam

Prostrate or low-climbing vine, twining at the tips. Stems unwinged, tomentose to glabrous; sap clear. **Leaves** alternate, simple, blade cordate, 9—27 x 5—25 cm, acuminate at the apex, cordate at the base; surfaces glabrous to pubescent; petiole 4—20 cm long. **Inflorescence** a compact, few-flowered axillary cyme, with a peduncle 5—20 cm long, and bearing several elliptic bracts 2—5.5 cm long. **Calyx** of 5 ovate sepals 2.5—4 cm long, on a pedicel 2.5—4.5 cm long. **Corolla** campanulate, shallowly 5-lobed, 5—9 cm long, white. Stamens 5, epipetalous; anthers coiled. Ovary superior. **Fruit** a membranous, translucent, subglobose capsule 2—2.5 cm in diameter, enclosed within the enlarged sepals, containing 4 black, glabrous seeds 7—12 mm in diameter.

Operculina ventricosa (Bertero) Peter is native to tropical America and was first recorded from the Pacific Islands in 1887 (Guam). It is occasional to locally common in disturbed lowland places, such as fallow fields and roadsides at up to 300 m elevation. It appears to have spread from Guam to other Pacific Islands by natural means (seawater-dispersed seeds), but was not recognized from Polynesia until recently.

This perennial vine can be distinguished by its unwinged stems, large, alternate, heart-shaped leaves, large (2—5.5 cm long) bracts, and showy white, bell-shaped corolla 5—9 cm long. It has often been mistaken for the widespread native species *Operculina turpethum* (L.) A. Manso, which has winged stems, smaller, acute-tipped leaves, and smaller bracts and seeds.

STICTOCARDIA TILIIFOLIA
Convolvulaceae (Morning-glory Family)

COMMON NAMES: pilikai (Hawai'i); pālulu, tagamimi (Samoa)
DISTRIBUTION: all the main island groups

Large perennial vine. Stems up to 10 m in length, finely pubescent, twining at the tip; sap clear. **Leaves** alternate, simple, blade cordate to ovate, 8—25 x 8—24 cm, shortly acuminate at the tip, cordate at the base; lower surface covered with fine black dots; margins entire; petiole 7—20 cm long. **Inflorescence** usually a 1—4-flowered cyme on a peduncle 2—20 cm long. **Calyx** 13—18 mm long, divided to near the base into 5 glabrous, nearly round sepals; pedicel 1.2—8 cm long. **Corolla** funnel-shaped, 7—10 cm long, shallowly 5-lobed, reddish purple with a darker center and radiating lines. Stamens 5, epipetalous. Ovary superior. **Fruit** a papery, globose capsule 2.5—3.5 cm in diameter, 4-seeded, enclosed by the large, papery sepals that disintegrate into a vascular framework.

Stictocardia tiliifolia (Desr.) Hall. f. is native from tropical Africa to Asia, but is now pantropic in distribution. It was first recorded from the Pacific Islands in 1864 (Hawai'i), where is occasional to locally common or abundant in disturbed places, such as thickets and fallow land, often climbing over low vegetation at up to 220 m elevation.

This perennial vine can be distinguished by its alternate, heart-shaped leaves with the lower surface marked with tiny black dots, axillary solitary flowers (or 2—4-flowered cymes), nearly round sepals, and large (7—10 cm long), funnel-shaped, light purple corolla darker in the center. Synonyms: *Ipomoea campanulata* of some authors, *Stictocardia campanulata* of some authors.

KALANCHOE PINNATA
Crassulaceae (Stonecrop Family)

COMMON NAMES: air plant; ‘oliwa ku kahakai (Hawai’i)
DISTRIBUTION: all the main islands (but rare in Samoa and Tonga)

Erect perennial herb up to 1.5 m in height. Stems hollow, glabrous, rooting at the base. **Leaves** opposite, succulent, some simple, mature ones often pinnately compound with 2—5 leaflets, simple blades elliptic, 10—25 x 5—16 cm, rounded at the apex, cuneate to acute at the base; surfaces glabrous; margins crenate, sometimes producing bulbils; petiole 2—4 cm long, purple. **Inflorescence** of terminal paniculate cymes 20—80 cm long bearing pendulous flowers. **Calyx** cylindrical, yellow streaked with red, 3—4 cm long, 4-lobed, inflated and papery. **Corolla** cylindrical, 4—5.5 cm long, divided less than 1/4 of its length into 5 ovate, red, acuminate-tipped lobes. Stamens 8, free. Ovaries 4, style as long at corolla. **Fruit** a narrowly ovoid follicle 1.2—1.8 cm long.

Kalanchoë pinnata (Lam.) Pers. is probably native to tropical Africa, but is now pantropic in distribution. It was first recorded from the Pacific Islands in 1871 (Hawai’i), where it is occasional to locally common in dry to moderately wet disturbed places, such as cultivated land, roadsides, and *Leucaena* scrub forest of Fiji and Hawai’i, at up to 550 m elevation. It is cultivated on many islands as an ornamental.

This large succulent herb can be distinguished by its simple and pinnately compound leaves often bearing marginal plantlets, large, drooping flowers with a tubular yellow calyx and red petals, and 4 ovaries. Synonyms: *Bryophyllum calycinum* Salisb., *B. pinnatum* (Lam.) Kurz. A related plant in Hawai’i, *Kalanchoë tubiflora* (Harv.) Raym.-Hamet, lacks the inflated calyx.

COCCINIA GRANDIS
Cucurbitaceae (Gourd Family)

COMMON NAMES: ivy gourd, scarlet-fruited gourd
DISTRIBUTION: Hawai'i, Tonga, Fiji, Guam

Dioecious perennial herbaceous vine. Stems mostly glabrous, produced annually from a tuberous rootstock; tendrils simple, axillary. **Leaves** alternate, simple, blade broadly ovate, 5-lobed, 5—9 x 4—9 cm, acute and mucronate at the apex, cordate with a broad sinus at the base; surfaces glabrous or scaly, with 3—8 glands near the base; margins denticulate; petiole 1—5 cm long. **Inflorescence** usually of solitary, axillary flowers. **Calyx** of 5 subulate, recurved lobes 2—5 mm long on the hypanthium; peduncle 1—5 cm long. **Corolla** campanulate, white, 3—4.5 cm long, deeply divided into 5 ovate lobes. Stamens 3, present as staminodes in female flowers. Ovary inferior. **Fruit** a smooth, bright red, ovoid to ellipsoid berry 2.5—6 cm long.

Coccinia grandis (L.) Voigt is native from tropical Africa to Australia and was first recorded from the Pacific Islands in 1940 (Fiji). It is locally abundant in the lowlands of Hawai'i (O'ahu and Kona Coast of the Big Island), where it climbs over low vegetation and even into tall trees, often entirely covering them. It has spread rapidly in Hawai'i since it was first noted in 1986 and is now a serious problem. Also naturalized in Fiji, but perhaps just an escape from cultivation in Tonga and Guam.

This herbaceous vine can be distinguished by its axillary tendrils, alternate, broadly ovate, 5-lobed leaves, showy white, bell-shaped flowers, and smooth red, ovoid to ellipsoid fruit up to 6 cm long.

MOMORDICA CHARANTIA
Cucurbitaceae (Gourd Family)

COMMON NAMES: balsam pear, bitter gourd
DISTRIBUTION: all the main island groups

Monoecious herbaceous vine. Stems sparingly pubescent; tendrils simple, axillary. **Leaves** alternate, simple, blade suborbicular, deeply palmately 5-lobed, 3—13 x 3—10 cm, acute at the apex, cordate with a broad sinus at the base; surfaces glabrous; margins irregularly toothed; petiole 1.5—10 cm long. **Inflorescence** of solitary, axillary flowers, on a peduncle 2—13 cm long bearing near the middle a cordate bract 3—20 mm long. **Calyx** deeply divided into 5 ovate lobes 4—6 mm long, on a hypanthium. **Corolla** of 5 obovate to spathulate, yellow petals 1—1.8 cm long. Stamens 3, present as staminodes in pistillate flowers. Ovary inferior, 1-celled. **Fruit** a fusiform to ovoid berry 3—12 cm long, orange with red pulp, longitudinally ridged and warty, containing several brown,ovoid seeds 8—16 mm long.

Momordica charantia L. is native to tropical or subtropical Asia or Africa and was first recorded from the Pacific Islands in 1864 (Fiji); a 1769 record (Tahiti) may be in error. It is locally common climbing over low vegetation in disturbed places such as thickets and fallow land, reported up to 300 m elevation. The fruits and young shoots are eaten, especially those of the larger-fruited, cultivated variety.

This herbaceous vine can be distinguished by its axillary tendrils, alternate, palmately lobed leaves, axillary, solitary, unisexual flowers on thin stalks bearing a heart-shaped bract near the middle, yellow corollas, and orange, spindle-shaped or ovoid fruits that split open to expose the red pulp and large brown seeds.

ACALYPHA LANCEOLATA
Spurge Family (Euphorbiaceae)

COMMON NAMES: ogogoo tea (Samoa)
DISTRIBUTION: Samoa, Tonga, Fiji, Belau

Erect monoecious herb 10—100 cm in height. Stems longitudinally grooved, appressed-pubescent. **Leaves** alternate, simple, blade ovate 1—8 x 0.8—5 cm, acute to acuminate at the apex, rounded at the base; margins coarsely serrate; surfaces densely pubescent, gland-dotted; petiole 0.5—5 cm long. **Inflorescence** of 1—4 axillary spikes 5—30 mm long. **Flowers** unisexual, apetalous, the lower ones female, upper ones male. Male flowers minute, less than 0.5 mm in diameter, clustered; stamens 8. Female flowers subtended by a pubescent, sheathing, cup-shaped bract 1.5—3 mm long with a serrate margin; styles 3, divided into filiform stigmas. **Fruit** a pubescent, 3-lobed schizocarp 1—2 mm long, splitting into 3 one-seeded segments.

Acalypha lanceolata Willd. is native to somewhere in the Old World tropics and was an ancient introduction as far eastward as Samoa or perhaps the Cook Islands. It is uncommon to occasional in disturbed places such as plantations, roadsides, and around houses.

This erect herb can be distinguished by its alternate, ovate leaves, toothed leaf margins, green, unisexual flowers on the same axillary spikes, female flowers enclosed within a cup-shaped bract with a toothed margin, and small fruit splitting into 3 one-seeded segments. Synonym: *Acalypha boehmeriodes* Miq. A similar species common in Samoa, Guam, and Belau, *Acalypha indica* L., differs in having longer petioles, an extra abnormal female flower on a long, terminal filament, and glabrous foliage.

CHAMAESYCE HIRTA
Euphorbiaceae (Spurge Family)

COMMON NAMES: garden spurge; koko kahiki (Hawai'i); 'āpulupulu (Samoa); sakisi (Tonga)
DISTRIBUTION: all the main island groups

Ascending, scarcely branched, monoecious herb 6—60 cm in height. Stems tomentose; sap milky. **Leaves** opposite, simple, blade ovate to lanceolate or oblong, mostly 1—5 x 0.5—2 cm, acute at the apex, acute to subcordate at the base; margins finely serrate; surfaces appressed-pubescent, often tinged or spotted red to purple; stipules linear; petiole 1—2 mm long. **Inflorescence** of 1 or 2 globose axillary cymes on a peduncle 3—15 mm long. **Flowers** in cyathia with a tiny 4- or 5-lobed involucre; appendage white or absent. Male flowers 2—8, minute, consisting of 1 stamen. Female flower 1, with a 3-lobed ovary; styles 3. **Fruit** a 3-lobed, pubescent, subglobose capsule 0.8—1.3 mm long, splitting into three 1-seeded segments; central column persistent.

Chamaesyce hirta (L.) Millsp. is probably native to tropical America, but is now pantropic in distribution. It was first recorded from the Pacific Islands in 1826 (Hawai'i), where it is common to locally abundant in disturbed places, such as roadsides and waste places, at up to 600 m elevation.

This herb can be distinguished by its hairy stems, milky sap, opposite leaves, tiny, inconspicuous cyathia (a type of inflorescence) arranged in globose cymes, and tiny, 3-lobed, splitting fruits. Synonyms: *Euphorbia hirta* L., *E. pilulifera* of some authors. Similar to the equally common *Chamaesyce hypericifolia* (see p. 69), which differs in being glabrous and having more conspicuous, mostly white or pink flowers.

CHAMAESYCE HYPERICIFOLIA
Euphorbiaceae (Spurge Family)

COMMON NAMES: graceful spurge
DISTRIBUTION: all the main island groups

Erect to ascending annual monoecious herb 10—60 cm in height. Stems glabrous; sap milky. **Leaves** opposite, simple, blade oblong to oblanceolate, 0.8—4 x 0.5—1.5 cm, round to subacute at the apex, oblique at the base; margins serrate; surfaces glabrous; stipules triangular; petiole 1—2 mm long. **Inflorescence** of 1 or 2 globose cymes on a peduncle 0.1—3 cm long, subtended by 2 tiny bracts. **Flowers** in cyathia with a tiny, glabrous, 4-lobed involucre; appendage white or pink, or absent. Male flowers 2—20, tiny, consisting of 1 stamen. Female flower 1, with a deeply lobed ovary; styles 3. **Fruit** a glabrous, 3-lobed, subglobose capsule 1—1.5 mm long, broadest near the middle, on a short stalk, splitting into three 1-seeded segments; central column persistent.

Chamaesyce hypericifolia (L.) Millsp. is native to tropical America, but is now widespread throughout the tropics. It was first recorded from the Pacific Islands in 1913 (Hawai'i), where it is occasional in disturbed lowland places such as roadsides, abandoned plantations, and waste areas.

This herb can be distinguished by its glabrous stems, milky sap, glabrous opposite leaves, tiny white flowers in axillary cyathia arranged above a pair of bracts, and tiny 3-lobed capsules. Synonyms: *Chamaesyce glomerifera* Millsp., *Euphorbia glomerifera* (Millsp.) Wheeler. Similar to *Chamaesyce hyssopifolia* (L.) Small of Hawai'i Samoa, and Fiji, which differs in having leafy cymes, narrower linear-oblong leaves, and a widely branching habit.

CHAMAESYCE PROSTRATA
Euphorbiaceae (Spurge Family)

COMMON NAMES: prostrate spurge
DISTRIBUTION: all the main island groups

Prostrate, mat-forming annual herb with a taproot. Stems 5—20 cm long, glabrous or finely pubescent on one side, purple; sap milky. **Leaves** opposite, simple, blade oblong to obovate, 0.3—0.8 x 0.15—0.5 cm, rounded at the apex, oblique to rounded at the base; margins serrate to subentire; surfaces glabrous; petiole 0.5—2 mm long. **Inflorescence** of few-flowered axillary cymes 2—4 mm long. **Flowers** in cyathia with a 4-lobed, subglabrous involucre 0.4—0.8 mm long; glands tiny, purple. Male flowers 2—5, tiny, consisting of 1 stamen. Female flower 1, consisting of a 3-lobed ovary; styles 3. **Fruit** an ovoid capsule 1—1.4 mm long, deeply 3-lobed, pilose on the angles, splitting into three 1-seeded segments; central column persistent.

Chamaesyce prostrata (left)
Chamaesyce thymifolia (right)

Chamaesyce prostrata (Aiton) Small is native to tropical America and was first recorded from the Pacific Islands in 1893 (Samoa). It is common in disturbed places, particularly around houses and in lawns, at up to 900 m elevation, but is apparently replaced in some places by the very similar *Chamaesyce thymifolia.*

This prostrate herb can be distinguished by its milky sap, purplish stems, small leaves with toothed margins, flowers in small axillary cyathia, and a small ovoid capsule with hairs on the 3 angles. Synonyms: *Euphorbia prostrata* Aiton. The similar *Chamaesyce thymifolia* (L.) Millsp., which is found in all the main island groups except Fiji, differs in having pink, pubescent stems, larger leaves, and ovoid capsules covered with hairs.

EUPHORBIA HETEROPHYLLA
Euphorbiaceae (Spurge Family)

COMMON NAMES: kaliko (Hawai'i)
DISTRIBUTION: Hawai'i, Samoa (rare), Guam

Erect annual monoecious herb 20-80 cm in height. Mostly glabrous; sap milky. **Leaves** alternate below, opposite above, simple, blade elliptic to obovate, 3—12 x 0.3—7 cm, acute to shortly acuminate at the apex, rounded to cuneate at the base; surfaces glabrous, glaucous; margins mostly entire; stipules absent or minute; petioles 1—4 cm long. **Inflorescence** a compact terminal cyme 1—2 long, subtended by white or green floral leaves. **Flowers** in cyathia, with a glabrous, 5-lobed involucre 2—2.5 mm long bearing a cup-shaped gland. Male flowers numerous, consisting of 1 stamen. Female flower 1, with a 3-lobed ovary; styles 3. **Fruit** a deeply 3-lobed, subglobose capsule 3—4 mm long, splitting to release the 3 gray to black or mottled seeds.

Euphorbia heterophylla L. is native to tropical or subtropical America and was first recorded in the Pacific Islands in 1871 (Hawai'i). It is occasional in dry, disturbed lowland places such as roadsides and waste places.

This erect herb can be distinguished by its milky sap, compact cymes above the green or white (but not red) floral leaves (bracts), unisexual flowers borne in cyathia, and deeply 3-lobed subglobose capsules splitting open to release the 3 seeds. Synonyms: *Euphorbia geniculata* Ort., *Poinsettia geniculata* (Ort.) Klotzsch & Garcke. A similar species present on most Pacific archipelagoes, *Euphorbia cyathophora* J.A. Murray, differs in having red floral bracts.

PHYLLANTHUS AMARUS
Euphorbiaceae (Spurge Family)

COMMON NAMES: none
DISTRIBUTION: all the main island groups except Hawai'i

Erect annual monoecious herb 10—60 cm in height. Stems sometimes woody at base. **Leaves** alternate, simple, reduced to scales on main stem, distichous on branchlets and appearing compound, blade oblong to oblanceolate, 0.4—1 x 0.2—0.5 cm, obtuse at the apex and base; margins entire; surfaces glabrous; stipules lanceolate; petiole *ca.* 0.5 mm long. **Inflorescence** of tiny, axillary, green, apetalous, unisexual flowers in mixed pairs. **Male flowers** with 5 reflexed calyx lobes *ca.* 0.5 mm long, margins scarious; pedicel 0.5—1.5 mm long. **Female flowers** with 5 oblong calyx lobes 0.6—0.9 mm long; margins broad, scarious; pedicel 1—2 mm long. **Fruit** a glabrous, green to yellow-brown, subglobose to 3-angled capsule 1.5—2 mm in diameter, splitting into six 1-seeded segments.

Phyllanthus amarus Schumach & Thonn. is native to tropical America and was first recorded from the Pacific Islands in 1847 (Tahiti). It is common in disturbed lowland places such as roadsides, waste places, and croplands in most of the islands where it occurs.

This erect, scarcely branching herb can be distinguished by its branches looking like compound leaves, small, simple, alternate leaves rounded at both ends, tiny green male and female flowers at the same nodes, and small green, globose fruits splitting into 6 sections. Synonym: *Phyllanthus niruri* of many authors. A similar species found in many islands (but rare in Hawai'i), *Phyllanthus urinaria* L., differs in having reddish stems, yellowish red, warty fruits, and acute leaf tips.

PHYLLANTHUS DEBILIS
Euphorbiaceae (Spurge Family)

COMMON NAMES: none
DISTRIBUTION: all the main island groups except Tonga

Erect, scarcely branching, annual monoecious herb 20—80 cm in height. Stems ridged, base sometimes woody. **Leaves** alternate, simple, reduced to scales on the main stem, distichous on the branchlets and appearing compound, blade narrowly elliptic to oblanceolate, 0.3—1.8 x 0.2—0.5 cm, subacute at the apex, acute at the base; margins entire; surfaces glabrous; petiole 0.3—1 mm long. **Inflorescence** of minute, green, apetalous, unisexual, axillary flowers. **Male flowers** in clusters of 2—4 in lower 2—4 nodes; calyx lobes 6, 0.5—1 mm long. **Female flowers** solitary at upper nodes; calyx lobes 6, obovate, 0.7—1.5 mm long; pedicels 1—2 mm long. **Fruit** a glabrous, subglobose capsule 1.8—2.3 mm in diameter, on a stalk 1—2 mm long, splitting into six 1-seeded segments.

Phyllanthus debilis (top)
Phyllanthus tenellus (bottom)

Phyllanthus debilis Klein ex Willd. is probably native to southern India or Sri Lanka and was first recorded from the Pacific Islands in 1871 (Hawai'i). It is common in disturbed lowland areas such as waste places, and occasionally on natural areas such as lava flows, at up to 250 m elevation.

This erect herb can be distinguished by it branches looking like compound leaves, simple, alternate, narrowly elliptic leaves acute at both ends, small male and female flowers at separate nodes, and subglobose capsules splitting into 6 sections. Synonyms: *Phyllanthus niruri* and *P. urinaria* of some authors in Hawai'i. A similar species found in Hawai'i and Tahiti, *Phyllanthus tenellus* Roxb., differs in having leaves broadly elliptic to obovate and fruit stalks 3—5 mm long.

RICINUS COMMUNIS
Euphorbiaceae (Spurge Family)

COMMON NAMES: castor bean; kolī (Hawai'i); lama pālagi (Samoa); lepo (Tonga)
DISTRIBUTION: all the main island groups

Shrub or small monoecious tree 1—4 m in height. Stems glabrous, often red. **Leaves** alternate, simple, blade suborbicular, peltate, up to over 40 cm across, deeply divided into 7—11 acute lobes; margins serrate; surfaces glabrous, often red; petiole about as long as blade. **Inflorescence** of narrow, erect, leaf-opposed cymose panicles up to 30 cm long, flowers apetalous, lower ones male, upper female. **Male flowers** 3—25 per cyme, calyx lobes 4 or 5, acute, 6—8 mm long; stamens hundreds, yellow. **Female flowers** in 1—7-flowered cymes, calyx lobes 4 or 5, caducous; styles 3, deeply bilobed, red; pedicel 25—40 mm long. **Fruit** a prickly, 3-lobed, ellipsoid to globose capsule 10—18 mm long, splitting into three 1-seeded segments; seeds mottled brown, 8—12 mm long.

Ricinus communis L. is native to Africa, but is now widely naturalized and cultivated in the tropics. It was first recorded from the Pacific Islands in 1819 (Hawai'i), where it is occasional in waste places, roadsides, and other disturbed, typically dry areas at up to 500 m elevation. Commercial castor oil is extracted from its poisonous seeds, and the leaves are often used in native remedies in the islands (Whistler 1992b).

This widely branching shrub can be distinguished by its often reddish foliage and stems, large, alternate, peltate, palmately lobed leaves, large panicles of unisexual flowers, the lower ones male and composed of numerous yellow stamens, the upper ones female and green, and prickly fruits that split into three 1-seeded segments.

ACACIA FARNESIANA
Fabaceae (Pea Family)

COMMON NAMES: klu, aroma; kolū (Hawai'i)
DISTRIBUTION: Hawai'i, Tahiti, Fiji, Guam

Spiny shrub up to 4 m in height. Young stems zigzag, mostly glabrous. **Leaves** alternate, bipinnately compound, pinnae in 2—7 pairs, rachis 2—8 cm long, with 1 or 2 nectaries below the lowest pinnae, leaflets in 8—25 pairs, blade oblong, 2.5—5 x 0.5—1 mm, acute at the apex, oblique and subsessile at the base; margins entire; surfaces glabrous; spines paired at the nodes, 4—40 mm long. **Inflorescence** mostly of solitary, axillary, globose heads 8—12 mm diameter, on a slender peduncle 2—3.5 cm long. **Calyx** tubular, 2.5—3.5 mm long, shortly 5-lobed. **Corolla** tubular, tiny, inconspicuous. Stamens yellow, numerous, exserted, 2—3 mm long. Ovary superior. **Fruit** a thick, straight or curved pod 5—7 x 1—2 cm, dark brown or black.

Acacia farnesiana (L.) Willd. is native to tropical America, but was introduced into many tropical countries because of the value of its flowers in making perfume. It was first recorded from the Pacific Islands in 1860 (Fiji), where it is occasional to common in dry, disturbed places, such as waste land and roadsides, especially in Hawai'i, at up to 400 m elevation.

This shrub can be distinguished by its sharp axillary spines, alternate, bipinnately compound leaves with small oblong leaflets, axillary globose heads bearing showy yellow stamens, and thick, curved or straight black pods.

ALYSICARPUS VAGINALIS
Fabaceae (Pea Family)

COMMON NAMES: alysicarpus
DISTRIBUTION: all the main island groups

Mostly prostrate annual herb. Stems pubescent when young, 10—60 cm long. **Leaves** alternate, appearing simple, leaflet blade obovate to oblong, 0.8—5 x 0.5—2 cm, rounded to subretuse at the apex, rounded to subcordate at the base; margins entire; lower surface with scattered hairs; stipule lanceolate, 3—15 mm long, persistent; petiole 3—12 mm long. **Inflorescence** a terminal raceme 1—4 cm long on a peduncle of variable length. **Calyx** 3—5.5 mm long, divided over halfway into 5 lanceolate lobes. **Corolla** papilionaceous, 5—6 mm long, purple to red with yellow markings. Stamens 10, diadelphous. Ovary superior. **Fruit** a compressed cylindrical pod 1—2.5 cm long, erect and clustered, black to yellow, breaking up into 1—8 one-seeded segments.

Alysicarpus vaginalis (L.) DC. is native to somewhere in the Old World tropics and was first recorded from the Pacific Islands in *ca.* 1900 (Fiji). It is occasional to common in lawns, on roadsides, and in other sunny, disturbed lowland places. It thrives in lawns because it can grow prostrate and withstand cutting or grazing.

This prostrate herb can be distinguished by its obovate to oblong leaflets appearing simple, short, terminal racemes of red to purple, pea-like flowers, and cylindrical pods breaking up into 1—8 segments. Synonym: *Alysicarpus nummularifolius* of some authors.

CALOPOGONIUM MUCUNOIDES
Fabaceae (Pea Family)

COMMON NAMES: calopo
DISTRIBUTION: all the main island groups except Hawai'i and Tonga

Creeping or trailing herb. Stems 1—4 m long, usually twining at the tips; all parts except corolla densely covered with long, tawny hairs. **Leaves** alternate, trifoliate, rachis 3—10 cm long, terminal blade ovate to rhomboid, 4—8 x 3—6 cm, rounded to acute at the apex, rounded at the base; lateral leaflets similar but smaller and unequally sided; margins entire; stipules ovate, 3—4 mm long. **Inflorescence** a few-flowered axillary raceme 6—12 cm long (in fruit). **Calyx** 5—8 mm long, divided to near the base into 5 linear-lanceolate lobes. **Corolla** papilionaceous, 7—10 mm long, light blue with a greenish or yellow blotch on the standard. Stamens 10, diadelphous. Ovary superior. **Fruit** a compressed, cylindrical pod 2—3 x 0.3—0.5 cm, containing 3—6 oblong brown seeds.

Calopogonium mucunoides Desv. is native to tropical America and was first recorded from the Pacific Islands in 1933 (Fiji). It was originally introduced as a cover crop, but has escaped in many places and is now locally common or abundant in disturbed places, such as roadsides, at up to 650 m elevation, especially in American Samoa. It grows rapidly, but is not very palatable to cattle.

This herbaceous vine can be distinguished by its twining stems, leaves and stems densely covered with yellow-brown hairs, trifoliate leaves, light blue, butterfly-like flowers in axillary racemes, and hairy cylindrical pods.

CHAMAECRISTA NICTITANS
Fabaceae (Pea Family)

COMMON NAMES: partridge pea; laukī (Hawai'i)
DISTRIBUTION: all the main island groups except Tahiti (rare in Samoa)

Erect, scarcely branching subshrub 50—150 cm in height. Stems red, pubescent. **Leaves** alternate, odd-pinnately compound, leaflets in 10—26 opposite pairs, rachis 3—8 cm long with a purple, flat-topped gland near the base, leaflet blade oblong, 1—2 x 0.12—0.25 cm, rounded and mucronate at the apex; midvein closer to one margin than the other; stipules lanceolate. **Inflorescence** of 1—3 flowers in short axillary racemes, each flower subtended by 2 persistent, linear-lanceolate bracteoles. **Calyx** 5—8 mm long, deeply divided into 5 linear-lanceolate lobes. **Corolla** 6—10 mm long, with 5 obovate yellow petals. Stamens 10, free, unequal in length. Ovary superior. **Fruit** a slightly curved, flattened, cylindrical pod 2.2—4.5 x 0.3—0.5 cm, red, pubescent.

Chamaecrista nictitans (L.) Moench is native to tropical America, but is now widespread throughout the tropics. It was first recorded from the Pacific Islands in 1871 (Hawai'i), where it is occasional to locally common in disturbed, dry to moderately wet places, such as roadsides and shrubby hillside vegetation, especially in Hawai'i, at up to 1100 m elevation.

This subshrub can be distinguished by its reddish stems, alternate, pinnately compound leaves with 10—26 pairs of narrowly oblong leaflets, yellow, 5-parted flowers, and flattened cylindrical pods 2.2—4.5 cm long. Synonyms: *Cassia lechenaultiana* DC., *C. mimosoides* of some authors, *Chamaecrista leschenaultiana* (DC.) Degener. A similar plant, *Chamaecrista mimosoides* (L.) Greene, has leaflets 2—8 mm long and narrower stipules.

CROTALARIA PALLIDA
Fabaceae (Pea Family)

COMMON NAMES: smooth rattlepod; pikakani (Hawai'i)
DISTRIBUTION: all the main island groups

Erect branching shrub 75—250 cm in height. Stems appressed-pubescent. **Leaves** alternate, trifoliate, rachis 1.5—6 cm long, blade obovate or oblanceolate to elliptic, 3—8 x 1—4 cm, acute to rounded and slightly notched at the apex, cuneate to rounded at the base; lower surface appressed-pubescent; stipules absent or tiny. **Inflorescence** a dense, terminal raceme 15—50 cm long, mostly 40—75-flowered. **Calyx** cup-shaped, 6—8 mm long, deeply divided into 5 acuminate lobes, appressed-pubescent. **Corolla** papilionaceous, 10—14 mm long, yellow with red veins, keel strongly curved. Stamens 10, diadelphous. Ovary superior. **Fruit** an inflated cylindrical pod 3.5—4.5 x 0.6—0.8 cm, subglabrous; seeds 20—55, kidney-shaped, 2.2—3.2 mm long, smooth.

Crotalaria pallida Aiton is possibly native to tropical Africa, but is now found throughout the tropics. It was first recorded from the Pacific Islands in 1864 (Hawai'i), where it is occasional in disturbed places such as pastures, waste places, and roadsides at up to 1100 m elevation.

This subshrub can be distinguished by its alternate, trifoliate leaves, long racemes of pea-like flowers, yellow petals marked with red lines, and inflated, nearly glabrous, cylindrical pods. Synonyms: *Crotalaria mucronata* Desv., *C. saltiana* of some authors, *C. striata* DC. A similar species found in Hawai'i, Guam, and elsewhere, *Crotalaria incana* L., differs in the absence of red striations on its petals and in the fuzzy pubescence of its leaves and pods.

DESMANTHUS VIRGATUS
Fabaceae (Pea Family)

COMMON NAMES: slender mimosa, virgate mimosa
DISTRIBUTION: all the main island groups except Tonga and Western Samoa

Scarcely branched subshrub up to 2 m in height. Stems strongly longitudinally ribbed. **Leaves** alternate, bipinnately compound, pinnae in 3—7 pairs, rachis 2—8 cm long with a red nectary between the lowest leaflet pair, leaflets in 10—25 pairs, blade linear, 1.5—6 x 0.5—2 mm, acute at the apex, oblique to subcordate at the base; stipules subulate, 2.5—8 mm long. **Inflorescence** in solitary, axillary heads 3—5 mm in diameter (excluding stamens), 6—10-flowered, on a stalk 1.5—6 cm long. **Calyx** campanulate, shallowly 5-lobed, 1.8—3 mm long. **Corolla** of 5 elliptic petals 3—4 mm long, white. Stamens usually 10, exserted, 4—6 mm long. Ovary superior. **Fruit** a linear pod 5—11 x 0.3—0.4 cm, usually 2—4 per head; seeds 15—35, 2—3 mm long, grayish brown.

Desmanthus virgatus (L.) Willd. is native to tropical America, but is now widespread in the tropics. It was first recorded from the Pacific Islands in 1912 (Hawai'i), where it is occasional as a weed of relatively dry, disturbed lowland places, especially in Hawai'i.

This erect, scarcely branching subshrub can be distinguished by its strongly ridged stems, alternate, bipinnately compound leaves, 3—7 pairs of pinnae, small linear leaflets, red nectaries on the leaf rachis, small, long-stalked heads of flowers, white, exserted stamens, and linear pods.

DESMODIUM INCANUM
Fabaceae (Pea Family)

COMMON NAMES: spanish clover; ka'imi (Hawai'i)
DISTRIBUTION: all the main island groups

Prostrate to erect subshrub 15—60 cm in height. Stems pubescent when young. **Leaves** alternate, trifoliate, rachis 1—4 cm long, blade elliptic to obovate, terminal one 1.5—7 x 0.8—3.5 cm, rounded to acute at the apex, rounded to subcordate at the base; upper surface with a lighter colored medial strip; stipule ovate to lanceolate. **Inflorescence** a terminal raceme 3—15 cm long with flowers in clusters of 1—3 subtended by 3 bracts. **Calyx** 1.4—2.8 mm long, divided into 5 narrow lobes; pedicel 3—8 mm long, persisting after fruit falls. **Corolla** papilionaceous, 4—6 mm long, pink to purple. Stamens 10, diadelphous. Ovary superior. **Fruit** a straight to slightly curved, fuzzy, sticky pod 2—3 cm long, one margin entire, the other notched into 3—6 oblong, one-seeded segments.

Desmodium incanum DC. is native to tropical America, but is now widespread in the tropics. It was first recorded from the Pacific Islands in 1916 (Hawai'i), where it is occasional to common in lawns, on roadsides, and in other sunny, disturbed areas at up to 460 m elevation.

This subshrub can be distinguished by its alternate, trifoliate leaves, elliptic to obovate leaflets, light-colored blotch along the midrib, racemes of small, purple to pink, pea-like flowers, and straight pod divided along one margin into 3—6 pubescent, oblong segments. Synonym: *Desmodium canum* (J. F. Gmelin) Schinz & Thell. Similar to *Desmodium sandwicense* E. Mey. of Hawai'i, which differs in having a pink to white corolla 8—10 mm long.

DESMODIUM TORTUOSUM
Fabaceae (Pea Family)

COMMON NAMES: Florida beggarweed
DISTRIBUTION: Hawai'i, Samoa, Tonga, Fiji, Guam

Erect, scarcely branching annual herb 30—200 cm in height. Stems densely pubescent. **Leaves** alternate, trifoliate, rachis 1.5—4 cm long, blade ovate to narrowly elliptic, terminal one 2—9 x 0.8—4 cm, apex and base rounded to acute; margins ciliate; surface subglabrous to pubescent; stipules ovate with an acuminate tip. **Inflorescence** a branching, terminal or axillary, many-flowered raceme 15—50 cm long, with solitary or paired flowers. **Calyx** 1.8—3 mm long, deeply lobed into 5 linear lobes, on a thin pedicel 8—16 mm long. **Corolla** papilionaceous, 4—6 mm long, pink or rarely yellow or white. Stamens 10, diadelphous. **Fruit** a straight, twisted pod 1.5—3 x 0.3—0.4 cm, margins equally deeply notched into 4—7 round, 1-seeded, pubescent segments.

Desmodium tortuosum (Sw.) DC. is native to tropical America, but is now widespread in the tropics. It was first recorded in the Pacific Islands in 1913 (Hawai'i), where it is occasional to locally common in pastures, on roadsides, and in other disturbed places at up to 670 m elevation.

This tall, erect subshrub can be distinguished by its trifoliate leaves, large, branching, many-flowered racemes, pink to white or yellow, pea-like flowers, and straight, twisted, papery pods notched on both margins into 4—7 round segments. Synonym: *Desmodium purpureum* of some authors.

DESMODIUM TRIFLORUM
Fabaceae (Pea Family)

COMMON NAMES: none
DISTRIBUTION: all the main island groups

Prostrate, creeping, much-branched herb. Stems pubescent, up to 50 cm long. **Leaves** alternate, trifoliate, rachis 3—8 mm long, blade obovate to obcordate, terminal one 4—10 x 3—10 mm, rounded to slightly notched at the apex, rounded to subcordate at the base; margins entire; surfaces appressed-pubescent; stipules ovate, acuminate at the tip, 2—6 mm long, strongly veined, persistent at the leafless nodes. **Inflorescence** of fascicles of 2—5 flowers opposite a leaf. **Calyx** 1.8—3.6 mm long, with 5 lanceolate lobes, on a pedicel 3—8 mm long. **Corolla** papilionaceous, 3—5 mm long, violet to purple. Stamens 10, diadelphous. Ovary superior. **Fruit** a flattened, slightly curved, membranous pod 6—18 x 2—3 mm, pubescent, notched on one margin into 3—5 one-seeded segments.

Desmodium triflorum (L.) DC. is native to somewhere in the Old World tropics, but is now pantropic in distribution. It was first recorded from the Pacific Islands in 1864 (Hawai'i), where it is occasional to locally common in sunny, disturbed places, such as lawns, roadsides, and lava flows at up to 600 m elevation.

This branching, prostrate herb can be distinguished by its trifoliate leaves, small obovate to obcordate leaflets, 2—5 purple, pea-like flowers in clusters opposite a leaf, and small, jointed papery pods notched along one margin into 3—5 segments. A similar species in Samoa, Tonga, Fiji, and Guam, *Desmodium heterophyllum* (Willd.) DC., differs in being a sprawling herb with larger elliptic leaves rounded at the tip, and longer pedicels.

INDIGOFERA SUFFRUTICOSA
Fabaceae (Pea Family)

COMMON NAMES: indigo; inikō (Hawai'i); 'akauveli (Tonga)
DISTRIBUTION: all the main island groups except Belau

Erect branching shrub 60—250 m in height. Stems longitudinally ridged, strigose, gray. **Leaves** alternate, odd-pinnately compound, rachis 5—11 cm long, strigose, leaflets in 4—8 opposite pairs, blade elliptic to obovate, rounded to acute and mucronate at the apex, rounded to cuneate at the base; margins entire; surfaces strigose, especially the lower side; stipule linear. **Inflorescence** a many-flowered axillary raceme 2—8 cm long. **Calyx** cup-shaped, 0.7—1.5 mm long, divided about halfway into 5 triangular lobes, strigose. **Corolla** papilionaceous, 3—5 mm long, salmon to pink. Stamens 10, diadelphous. Ovary superior, 1-celled. **Fruit** cylindrical, curved, 12—20 x 2—3 mm, appressed pubescent, 4—6 seeded; seeds cylindrical, *ca.* 2 mm long, dark brown to black.

Indigofera suffruticosa Mill. is native to tropical America, but is now pantropic in distribution. It was first recorded from the Pacific Islands in 1836 (Hawai'i), where it is occasional to locally common in disturbed places such as roadsides, croplands, and waste places, at up to 1100 m elevation. It was originally introduced to some islands for use as a dye plant.

This shrub can be distinguished by its grayish foliage, pinnately compound leaves, short axillary racemes of small, salmon-red, pea-like flowers, and short, curved, cylindrical pods. Synonym: *Indigofera anil* L. A similar but smaller species found in Hawai'i, Tahiti, Fiji, and Micronesia, *Indigofera spicata* Forssk., differs in having straight, longer pods and typically being prostrate.

LEUCAENA LEUCOCEPHALA
Fabaceae (Pea Family)

COMMON NAMES: wild tamarind; koa haole, haole koa (Hawai'i); fuapepe (Samoa); siale mohemohe (Tonga); nito (Cook Islands); vaivai (Fiji); tangantangan (Guam)
DISTRIBUTION: all the main island groups

Woody shrub or small tree 1.5—5 m or more in height. **Leaves** alternate, bipinnately compound, pinnae in 3—8 opposite pairs, rachis mostly 8—25 cm long with a gland near the lowest pair of pinnae, leaflets in 8—20 subopposite pairs, blade linear-lanceolate, unequally sided, 8—18 x 3—5 mm, acute at the apex, oblique and subsessile at the base; surfaces glabrous; stipules subulate. **Inflorescence** of 1—3 dense, axillary, globose heads 1—1.8 cm in diameter, on a peduncle 2.5—6 cm long that becomes woody. **Calyx** campanulate, 2—3.2 mm long, shallowly 5-lobed. **Corolla** of 5 lanceolate petals 4—5 mm long, white. Stamens 10, exserted, 6—8 mm long, showy white. Ovary superior. **Fruit** a flat, membranous, strap-shaped pod 10—20 x 1.5—2 cm, dark brown, containing 15—25 shiny brown seeds 6—9 mm long.

Leucaena leucocephala (Lam.) de Wit is native to tropical America, perhaps to the Caribbean Islands, and was first recorded from the Pacific Islands in 1837 (Hawai'i). It is common to abundant in dry areas at up to 800 m elevation, often dominating these habitats. It is an important firewood in some places, and the seeds are strung into leis. A psyllid insect species devastated the island populations in the 1980s.

This small tree can be distinguished by its alternate, bipinnately compound leaves, small, linear-lanceolate leaflets, globose heads of white flowers with exserted stamens, and clusters of strap-shaped pods borne on a woody stalk. Synonym: *Leucaena glauca* (L. ex Willd.) Benth.

MACROPTILIUM LATHYROIDES
Fabaceae (Pea Family)

COMMON NAMES: cow pea, phasey bean
DISTRIBUTION: Hawai'i, Samoa, Tonga, Tahiti, Fiji

Erect herb up to 1 m or more in height. Stems finely pubescent, longitudinally grooved, somewhat twining at the tip. **Leaves** alternate, trifoliate, rachis mostly 1.5—5 cm long, blade ovate to oblong, 2—6 x 1—3.5 cm, acute and mucronate at the apex, cuneate at the base; surfaces appressed-pubescent; margins entire; stipules lanceolate, 4—6 mm long. **Inflorescence** of several-flowered axillary racemes 25—60 cm long. **Calyx** cylindrical, 4—7 mm long, longitudinally ribbed, shallowly lobed into 5 acute, subequal teeth. **Corolla** papilionaceous, dark maroon, 12—15 mm long. Stamens 10, diadelphous. Ovary superior. **Fruit** a brown, pubescent, linear, upward-curved pod 7—12 cm long, twisting and splitting open. Seeds 18—30, compressed-oblong, 2.5—3.5 mm long.

Macroptilium lathyroides (L.) Urb. is native to tropical America and was first recorded from the Pacific Islands in 1864 (Hawai'i). It is occasional to common in disturbed places, such as roadsides and pastures, at up to 600 m elevation.

This erect herb can be distinguished by its somewhat twining stem tips, alternate, trifoliate leaves, ovate to oblong leaflets, dark maroon, pea-like flowers, and linear pods that split open into twisted valves. Synonyms: *Phaseolus lathyroides* L., *P. semierectus* L. A similar species on all the main islands except Belau, *Macroptilium atropurpureum* (DC.) Urb., is a vine with broader, typically lobed leaflets densely pubescent on the lower surface and darker maroon flowers.

MIMOSA INVISA
Fabaceae (Pea Family)

COMMON NAMES: giant sensitive plant; vao fefe pālagi (Samoa)
DISTRIBUTION: Samoa, Tahiti, Fiji, Belau

Spreading shrub forming tangled masses up to 1.5 m in height, with leaves somewhat sensitive to the touch. Stems hispid, prickly. **Leaves** alternate, bipinnately compound, rachis prickly, 7—15 cm long, pinnae in 3—8 opposite pairs 2.5—5 cm long, leaflets in 12—25 opposite pairs, blade narrowly oblong, 5—10 x 1—2 mm, broadly acute to round at the apex, obliquely rounded and subsessile at the base; surfaces pubescent; stipules subulate, hispid on margins. **Inflorescence** of solitary, axillary heads 4—8 mm in diameter, on a prickly peduncle 5—12 cm long. **Calyx** tiny. **Corolla** 1.5—2.5 mm long, 4-lobed, pink. Stamens 8, pink, 3—6 mm long. Ovary superior. **Fruit** a flat, oblong, bristly pod 1—3 cm long, splitting into 2—5 one-seeded segments.

Mimosa invisa Mart. ex Colla is native to tropical America, but is now widely distributed in the tropics. It is a recent introduction to the islands, where it was first recorded in 1936 (Fiji), and is occasional but tends to dominate where it does occur in disturbed areas such as plantations and roadsides. It is reported from near sea level to 600 m elevation. Because of its nasty prickles and scrambling habit, it is a noxious weed.

This scrambling shrub can be distinguished by its prickly stems and leaf rachises, alternate, bipinnately compound leaves, 3—8 pairs of pinnae, solitary, axillary, stalked heads of pink flowers with exserted stamens, and bristly pods that break up into sections.

MIMOSA PUDICA
Fabaceae (Pea Family)

COMMON NAMES: sensitive plant, sleeping grass; vao fefe (Samoa); mateloi (Tonga); pohe ha'avare (Tahiti); co gadrogadro (Fiji)
DISTRIBUTION: all the main island groups

Low, decumbent, loosely-branching subshrub up to 40 cm in height, with leaves closing up when touched. Stems reddish, with scattered, curved prickles. **Leaves** alternate, bipinnately compound, rachis 1.5—5 cm long, pinnae terminal in 1 or 2 pairs, 2—6 cm long, leaflets in 6—20 pairs, blade narrowly oblong, 4—10 x 1.2—3 mm, acute at the apex, obliquely round at the base; surfaces appressed-hairy; stipule lanceolate with bristly margins. **Inflorescence** of solitary, axillary, ovoid to globose heads 6—12 mm across, on a peduncle 1—4 cm long. **Calyx** tiny. **Corolla** 1.7—2.5 mm long, 4-lobed, pink. Stamens 4, 4—7 mm long, pink to purple. Ovary superior. **Fruit** a flat, oblong, somewhat bristly pod 8—15 mm long, splitting into 2—4 one-seeded segments with entire margins.

Mimosa pudica L. is native to tropical America, but is now pantropic in distribution. It was first recorded from the Pacific Islands in 1839 (Samoa), where it is common to locally abundant in disturbed sunny places, especially in lawns and pastures, at up to 700 m elevation. It can be a serious pest because of its prickles, which hamper hand-weeding.

This subshrub can be distinguished by its reddish, prickly stems, alternate, bipinnately compound leaves, 1—2 pairs of terminal pinnae, solitary, axillary, stalked heads of pink flowers with exserted stamens, and 2—4-segmented pods with bristly margins. It differs from the larger *Mimosa invisa* (p. 87), which has 3—8 pairs of pinnae equally spaced on the rachis.

PITHECELLOBIUM DULCE
Fabaceae (Pea Family)

COMMON NAMES: Manila tamarind, Madras thorn; ‘opiuma (Hawai’i); kamachile (Guam)

DISTRIBUTION: Hawai’i (but cultivated elsewhere)

Tall spiny tree up to 15 m in height. Stems marked with white lenticels. **Leaves** alternate, bipinnately compound, pinnae in 1 pair, rachis 1—5 cm long, with a nectary between the pinnae, leaflets 1 pair per pinna, blade elliptic, unequally sided, 1—5.5 x 0.5—2.5 cm, rounded at the apex and base; margins entire; surfaces glabrous; spines paired at the nodes. **Inflorescence** of axillary racemes or panicles of heads 6—9 mm in diameter. **Calyx** campanulate, 0.5—1 mm long, shallowly 4—6-lobed. **Corolla** funnelform, 4—6-lobed, 2.5—4 mm long, white. Stamens many, exserted, white. Ovary superior. **Fruit** a slightly flattened, spirally twisted cylindrical pod 10—15 x 1—1.6 cm, brown to red or pink, containing 3—8 obovate, glossy black seeds embedded in a white to red aril.

Pithecellobium dulce (Roxb.) Benth. is native to tropical America from Mexico to Venezuela and was first recorded from the Pacific Islands in 1871 (Hawai’i). It is cultivated as an ornamental tree around villages and along streets on some islands, but is naturalized in Hawai’i in dry scrubland and disturbed forest of the lowlands. The aril around the seeds is sometimes eaten.

This large tree can be distinguished by its paired spines at the nodes, alternate, bipinnately compound leaves with 4 unequally sided leaflets, panicles or racemes of small globose heads, white flowers, and twisted pods containing several shiny black seeds embedded in an aril.

PROSOPIS PALLIDA
Fabaceae (Pea Family)

COMMON NAMES: mesquite; kiawe, algaroba (Hawai'i)
DISTRIBUTION: Hawai'i, cultivated in Fiji and Guam

Spiny tree up 20 m in height. Stems glabrous. **Leaves** alternate, bipinnately compound, pinnae in 2—4 pairs, rachis 1.2—4 cm long with a nectary below each pair of pinnae, leaflets in 6—15 opposite pairs, blade oblong, 2.5—9 x 1.4—3 mm, rounded to obtuse at the apex, obliquely round and subsessile at the base; margins entire, ciliate; surfaces mostly glabrous, veins prominent on lower side; spines paired at nodes, 3—20 mm long. **Inflorescence** a dense, many-flowered, cylindrical, axillary, solitary spike 7—14 cm long. **Calyx** cup-shaped, *ca.* 0.5 mm long, shallowly 5-lobed. **Corolla** of 5 petals 2.5—3 mm long, yellow-green. Stamens 10, exserted, white, 4—6 mm long. Ovary superior. **Fruit** a subcylindrical, curved or straight pod 8—18 x 1—1.5 cm, on a stalk 1—1.8 cm long.

Prosopis pallida (Humb. & Bonpl. ex Willd.) Kunth is native to tropical South America (Peru, Columbia, and Ecuador) and was first recorded from the Pacific Islands in 1828 (Hawai'i). It is common and often forms monodominant forests in the dry, disturbed coastal areas and lowlands of Hawai'i, sometimes extending up to 600 m elevation.

This tree can be distinguished by its axillary spines, alternate, bipinnately compound leaves, axillary cylindrical spikes, tiny white flowers, exserted stamens, and yellowish brown subcylindrical pods. Synonyms: *Prosopis chilensis* and *P. juliflora* of some Hawaiian authors. When young, it is similar to *Acacia farnesiana* (see p.75), but differs in having a gland by each pair of pinnae.

PUERARIA LOBATA
Fabaceae (Pea Family)

COMMON NAMES: kudzu; a'a (Samoa); aka (Tonga); yaka (Fiji)
DISTRIBUTION: all the main island groups

Twining or creeping vine with a tuberous root. Stems densely covered with yellow-brown hairs. **Leaves** alternate, trifoliate, rachis 8—20 cm long, terminal blade mostly ovate and lobed, 8—22 x 7—15 cm, shortly acuminate at the apex, rounded at the base; lateral leaflets similar, but irregularly lobed; surfaces softly pubescent; stipules produced above and below their attachment. **Inflorescence** a many-flowered axillary raceme 15—40 cm long. **Calyx** 9—14 mm long, divided more than halfway into 5 unequal lobes, hairy, subtended by a pair of small bracteoles. **Corolla** papilionaceous, 13—20 mm long, violet with the standard marked with a yellow blotch. Stamens 10, diadelphous. Ovary superior. **Fruit** a flat, densely hairy pod 9—12 x 0.9—1.2 cm, rarely seen.

Pueraria lobata (Willd.) Ohwi is native to eastern or southern Asia, but was an ancient introduction from India to Samoa, and a European introduction to Hawai'i, the Society Islands, and Guam. It is occasional to locally common as a weed of disturbed places, especially in Samoa on roadsides and in abandoned plantations, and is reported up to 400 m elevation. The cooked roots were formerly eaten as a famine food in Polynesia and elsewhere.

This vine can be distinguished by its stems covered with yellow-brown hairs, large, pubescent, alternate, trifoliate leaves, long racemes of pea-like flowers, and purple corolla with a yellow blotch. Synonyms: *Pueraria harmsii* Rechinger, *P. thunbergiana* (Sieg. & Zucc.) Benth.

SAMANEA SAMAN
Fabaceae (Pea Family)

COMMON NAMES: monkeypod tree, rain tree; ‘ohai (Hawai’i)
DISTRIBUTION: all the main island groups (but only cultivated in most)

Large tree up to 25 m in height with a spreading crown. Stems puberulent when young, bark gray, rough, furrowed. **Leaves** alternate, bipinnately compound, pinnae in 4—7 pairs 15—25 cm long with a nectary near each leaflet pair, leaflets 3—9 pairs per pinna, blade obliquely obovate with a diagonal midrib, 1.5—6 x 0.7—3 cm, rounded at the apex, oblique at the base, subsessile; upper surface glossy dark green, lower puberulent; margins entire. **Inflorescence** of 1—4 axillary heads on peduncles mostly 5—10 cm long. **Calyx** narrowly campanulate, 6—7.5 mm long, puberulent, 5-toothed. **Corolla** funnelform, 9—13 mm long, pink with 5 green lobes. Stamens many, showy, up to 3 cm long. Ovary superior. **Fruit** a black or dark brown, sausage-shaped pod 15—20 x 1.5—2.5 cm, with thickened edges.

Samanea saman (Jacq.) Merr. is native from Mexico to Brazil and Peru, and was first recorded from the Pacific Islands in 1871 (Hawai’i). It is commonly planted as an ornamental tree along streets and in yards, persisting in old plantations and somewhat naturalized in other disturbed lowland areas, at least in Hawai’i, Fiji, and Samoa, extending up to 300 m elevation.

This large spreading tree can be distinguished by its alternate, bipinnately compound leaves, 4—7 pairs of pinnae, 3—9 pairs of unequally sided, dark green obovate leaflets per pinna, flowers in clusters with long pink stamens, and dark brown or black, sausage-shaped pods. Synonym: *Pithecellobium saman* (Jacq.) Benth., the name used in the flora of Fiji (Smith 1985).

SENNA OCCIDENTALIS
Fabaceae (Pea Family)

COMMON NAMES: coffee senna; mikipalaoa (Hawai'i)
DISTRIBUTION: all the main island groups

Erect subshrub 1—2 m in height. Stems mostly glabrous, longitudinally grooved. **Leaves** alternate, odd-pinnately compound, leaflets in 3—6 opposite pairs, rachis 8—13 cm long with a globose gland at the base, blade ovate to lanceolate, 2.5—10 x 1.5—3 cm, acute to acuminate at the apex, obliquely rounded at the base; surfaces mostly glabrous; stipule subulate, caducous. **Inflorescence** a terminal or axillary raceme 1—2 cm long, 2—4-flowered with lanceolate bracts 1—2 cm long. **Calyx** 7—12 mm long, divided to near the base into 5 obovate sepals. **Corolla** of 5 petals 9—14 mm long, yellow; fertile stamens 6, unequal in size; ovary superior. **Fruit** a glabrous, slightly curved, flattened cylindrical pod 9—15 x 0.7—1 cm, with broad margins lighter in color.

Senna occidentalis (L.) Link is an early modern introduction to the islands, where it was first recorded in 1839 (Samoa). It is probably native to tropical America, but now is pantropical in distribution, and is occasional to common in disturbed places such as roadsides, croplands, and pastures, at up to 1300 m elevation. It is sometimes used as a substitute for coffee.

This shrub can be distinguished by its alternate, pinnately compound leaves, 3—6 pairs of ovate to lanceolate leaflets, short racemes of yellow, 5-parted flowers, and long, flattened-cylindrical pods with lighter colored margins. Synonym: *Cassia occidentalis* L. Another widespread species (but not in Hawai'i or Belau), *Senna tora* (L.) Roxb., differs in having bad-smelling foliage, leaflets round at the apex, and slender pods 4—6 mm in diameter.

SENNA SURATTENSIS
Fabaceae (Pea Family)

COMMON NAMES: kolomona, kalamona (Hawai'i)
DISTRIBUTION: Hawai'i, cultivated in Tahiti and Guam

Small tree up to 6 m in height. Stems glabrous. **Leaves** alternate, odd-pinnately compound, leaflets 6—10 pairs, rachis 6—12 cm long with clavate nectaries between the first 1—3 leaflet pairs, blade obovate or oblanceolate to elliptic, 1.5—5 x 0.8—2 cm, rounded at the apex and base; margins entire; upper surface glabrous, lower glaucous; stipules linear. **Inflorescence** of axillary, few-flowered racemes on a peduncle 3—10 cm long, with caducous, ovate to lanceolate bracts. **Calyx** deeply divided into 5 unequal, round to oblanceolate sepals 4—8 mm long. **Corolla** of 5 ovate to oblong, yellow, unequal petals 12—24 mm long. Stamens 10, subequal. Ovary superior. **Fruit** a pendulous, strap-shaped, membranous pod 5—12 x 1—1.5 cm.

Senna surattensis (Burm.) H. Irwin & Barneby is native to Australia or Southeast Asia, but is now widely cultivated in the tropics, where it sometimes becomes naturalized. It was first recorded from the Pacific Islands in 1871 (Hawai'i), where it is occasional to common in disturbed lowland scrub forests in Hawai'i, and is cultivated there and on other islands.

This small tree can be distinguished by its alternate, pinnately compound leaves, 6—10 pairs of glaucous leaflets, yellow flowers, and flattened pods. Synonyms: *Cassia glauca* of Hawaiian authors, *C. surattensis* Burm., *Psilorhegma glauca* of Degener. A similar species naturalized in Hawai'i, *Senna pendula* (Humb. & Bonpl. ex Willd.) H. Irwin & Barneby, differs in having fewer fertile stamens and hanging, cylindrical pods.

HYPTIS PECTINATA
Lamiaceae (Mint Family)

COMMON NAMES: comb hyptis
DISTRIBUTION: all the main island groups

Narrow erect perennial herb or subshrub up to 3 m in height. Stems often woody at the base, finely pubescent, 4-angled. **Leaves** opposite, simple, blade mostly ovate, 1.5—5 x 1—3 cm, mostly acute at the apex, rounded to nearly truncate at the base; upper surface nearly glabrous, lower densely pubescent; margins crenate; petiole 0.5—3 cm long. **Inflorescence** of dense, whorl-like, many-flowered clusters on axillary cymes, or these in panicles mostly 4—20 cm long. **Calyx** tubular, 2—5 mm long including the 5 linear lobes, 10-ribbed, pubescent, sessile. **Corolla** bilabiate, white to pale violet, 2.5—3.5 mm long. Stamens in 2 pairs, epipetalous. Ovary superior. **Fruit** consisting of 4 nutlets 1—1.5 mm long, enclosed within the persistent calyx.

Hyptis pectinata (L.) Poit. is native to tropical America, but is now widespread in the tropics. It was first recorded from the Pacific Islands before 1819 (Guam), where it is occasional to locally abundant in disturbed places such as roadsides, fallow land, and pastures at up to 550 m elevation. It is a serious pest, particularly in pastures, since cattle do not eat it.

This tall herb can be distinguished by its square stems, opposite, ovate leaves, crenate leaf margins, axillary cymes or panicles of cymes, tiny tubular calyces with 5 linear teeth, tiny white to pale violet, 2-lipped flowers, and fruit consisting of 4 nutlets enclosed within the persistent calyx. A similar weedy and cultivated species from Hawai'i and Guam, *Hyptis suaveolens* (L.) Poit. differs in having a larger calyx 5—10 mm long.

HYPTIS RHOMBOIDEA
Lamiaceae (Mint Family)

COMMON NAMES: vao mini (Samoa); botones (Guam)
DISTRIBUTION: Samoa, Tahiti, Guam, Belau

Erect herb up to 2 m in height. Stems 4-angled. **Leaves** opposite, simple, ovate to lanceolate, 5—25 x 1—13 cm, acute to attenuate at the apex, cuneate at the base; upper surface hispid, lower glandular-punctate with pubescent veins; margins coarsely and irregularly serrate; petiole 0.5—6 cm long, narrowly winged. **Inflorescence** a subglobose, axillary, many-flowered head 1—3 cm in diameter, subtended by 5—10 lanceolate to narrowly oblong, unequal bracts; peduncle 1—6 cm long. **Calyx** campanulate, 3.5—5 mm long at anthesis including the 5 linear lobes, sessile. **Corolla** bilabiate, 5—5.5 mm long, white. Stamens 4, epipetalous. Ovary superior. **Fruit** consisting of 4 dark, ellipsoid nutlets 1—1.5 mm long, enclosed within the persistent calyx up to 7 mm long.

Hyptis rhomboidea Mart. & Gal. is native to tropical America and was first recorded from the Pacific Islands prior to 1970 (Guam). It is common in disturbed places such as pastures, fallow land, and savannas (Guam) in the lowlands. It is a recent (first recorded in 1979) and rapidly spreading noxious weed in Western Samoa.

This robust herb can be distinguished by its square stems, irregularly toothed, opposite leaves, long-stalked, globose, axillary heads, white, 2-lipped flowers, and fruits consisting of 4 nutlets. It is often mistakenly called *Hyptis captitata* Jacq., which apparently is not found in the islands.

LEONOTIS NEPETIFOLIA
Lamiaceae (Mint Family)

COMMON NAMES: lion's ear
DISTRIBUTION: Hawai'i

Coarse annual herb mostly 0.8—1.6 m in height. Stems 4-angled, puberulent. **Leaves** opposite, simple, oblong to ovate, 4—9 x 1.5—5 cm, acute to acuminate at the apex, cuneate to acuminate at the base; surfaces puberulent; margins coarsely crenate to serrate; petiole 2—7.5 cm long. **Inflorescence** a dense, globose structure formed by a sessile cyme from each axil, 2.5—6 cm in diameter. **Calyx** narrowly funnel-shaped, 1.2—2 mm long, shallowly divided into 8 or 9 unequal, spinulose teeth, sessile. **Corolla** bilabiate, tubular, somewhat curved, 2—2.8 cm long, orange. Stamens 4, epipetalous, in unequal pairs. Ovary superior. **Fruit** of 4 dark, oblong to obovoid nutlets 2.5—4 mm long, enclosed within the persistent calyx that is up to 2.3 cm long.

Leonotis nepetifolia (L.) R. Br. is native to tropical Africa and was first recorded in the Pacific Islands in 1938 (Hawai'i). It is occasional to common in disturbed places such as roadsides and cultivated land at up to 300 m elevation. It was originally introduced as an ornamental and for use in dry flower arrangements.

This erect herb can be distinguished by its square stems, opposite, coarsely toothed leaves, dense globose clusters of flowers surrounding the stems at the axils, spine-tipped calyces, showy orange, 2-lipped tubular corollas, and fruits composed of 4 dark nutlets.

OCIMUM GRATISSIMUM
Lamiaceae (Mint Family)

COMMON NAMES: wild basil
DISTRIBUTION: Hawai'i, Samoa, Tahiti

Erect perennial herb mostly 1—2 m in height. Stems 4-angled, pubescent when young. **Leaves** opposite, simple, strongly fragrant, blade mostly ovate, 3—10 x 1.5—8 cm, acuminate at the apex, cuneate at the base; surfaces glandular-punctate, glabrous to sparsely puberulent on the veins; margins serrate; petiole 1—6 cm long. **Inflorescence** of terminal and axillary racemes mostly 7—20 cm long, with flowers arranged in whorls. **Calyx** bilabiate, campanulate, 10-nerved, 2.5—4 mm long at flowering, lower lobe 4-toothed; pedicel 2—3 mm long. **Corolla** bilabiate, campanulate, 4—7 mm long, greenish white to yellowish. Stamens in 2 pairs, epipetalous, exserted. Ovary superior. **Fruit** of 4 subglobose nutlets 1.5—2 mm long, enclosed within the enlarged membranous calyx up to 7 mm long.

Ocimum gratissimum L. is native to somewhere in the Old World tropics, but is now pantropic in distribution. It was first recorded from the Pacific Islands in 1924 (Hawai'i), where it is common in disturbed places, especially in *kiawe* forests on Hawai'i, at up to 400 m elevation.

This robust herb can be distinguished by its toothed, opposite, fragrant leaves, racemes bearing whorls of flowers, greenish, 2-lipped flowers, and fruits of 4 nutlets in the membranous, 2-lipped calyx. Synonym: *Ocimum suave* Willd. The fuzzy variety *suave* (Willd.) Hook. is found in Samoa, Tahiti, and the Cook Islands. *Ocimum basilicum* L. (sweet basil) is mostly cultivated and has an open, rather than a closed, fruiting calyx. The smaller *Ocimum tenuiflorum* L. is semi-naturalized in Samoa and elsewhere.

CUPHEA CARTHAGENENSIS
Lythraceae (Loosestrife Family)

COMMON NAMES: tarweed
DISTRIBUTION: all the main islands except Guam and Belau

Erect branching herb 10—60 cm in height. Stems glandular-pubescent, sticky, reddish. **Leaves** opposite, simple, blade elliptic to obovate, 1—5 x 0.6—2.2 cm, acute to nearly obtuse at the apex, acute and decurrent at the base; surfaces scabrous; margins entire; petiole 0—2 mm long. **Inflorescence** of sessile, axillary, usually solitary flowers. **Calyx** a tubular hypanthium 4.5—6.5 mm long, 12-ribbed, shallowly divided into 6 minute teeth less than 0.5 mm long. **Corolla** of 6 slightly unequal, pink or purple petals 1.5—2 mm long. Stamens 11, free, attached to the calyx tube. Ovary superior. **Fruit** an ovoid capsule 3—4 mm long, surrounded by the inflated hypanthium, and containing 4—8 brown, compressed-suborbicular seeds 1.3—1.8 mm in diameter.

Cuphea carthagenensis (Jacq.) Macbr. is native to South America and was first recorded from the Pacific Islands in 1851 (Hawai'i). It is occasional to locally common in relatively wet, disturbed places such as pastures and roadsides at up to 950 m elevation.

This herb can be distinguished by its sticky, reddish stems, opposite leaves, flowers solitary or few in the axils, 6 small purple petals attached at the top of the calyx tube, and small ovoid capsules enclosed within the distinctly 12-ribbed calyx tube. Synonyms: *Cuphea balsamona* (Vand.) Cham. & Schlecht., *C. hyssopifolia* of Hillebrand, not Kunth.

ABUTILON GRANDIFOLIUM
Malvaceae (Mallow Family)

COMMON NAMES: hairy abutilon; ma'o (Hawai'i)
DISTRIBUTION: Hawai'i and Tahiti

Erect shrub 1—3 m in height. Stems densely hirsute with hairs up to 5 mm long. **Leaves** alternate, simple, blade ovate to round, 4—24 x 2.5—20 cm, subacuminate at the apex, deeply cordate at the base; upper surface subglabrous, lower densely stellate-pubescent; margins dentate; stipules linear-lanceolate; petioles 2—15 cm long. **Inflorescence** axillary, flowers solitary or in short, 2—6-flowered cymes, on a peduncle 2—10 cm long. **Calyx** 1.1—2 cm long, lobed to near the base into 5 ovate sepals. **Corolla** cup-shaped to nearly rotate, with 5 yellow-orange petals 1.2—2.5 cm long, lobed apically. Stamens numerous, monadelphous. Ovary superior, stigma red. **Fruit** a dull black, subglobose, deeply lobed, pubescent schizocarp 13—25 mm in diameter, splitting into 10 short-beaked segments.

Abutilon grandifolium (Willd.) Sweet is native to tropical America and was first recorded from the Pacific Islands in 1903 (Hawai'i). It is occasional to common in disturbed places such as roadsides, waste areas, and *Prosopis* (kiawe) forest at up to 1000 m elevation. It is sometimes cultivated as an ornamental or for its stem fibers.

This shrub can be distinguished by it densely hairy stems, alternate, ovate to round leaves, toothed leaf margins, yellow-orange flowers, stamens fused into a column, and black, hairy, splitting fruits. Synonym: *Abutilon molle* (Ort.) Sweet. A similar species rare or uncommon in Hawai'i, Tahiti, and Guam, *Abutilon indicum* (L.) Sweet, differs in having larger flowers (2.5—3 cm long) and more fruit segments (15—22).

MALVA PARVIFLORA
Malvaceae (Mallow Family)

COMMON NAMES: cheeseweed
DISTRIBUTION: Hawai'i

Annual or short-lived perennial, decumbent to erect herb 20—50 cm in height. Stems stellate-pubescent. **Leaves** alternate, simple, blade reniform to round, shallowly 3—7-lobed, mostly 1.5—8 x 1—8 cm, rounded at the apex, cordate at the base; surfaces stellate-pubescent; margins wavy to crenate; stipules ovate to lanceolate; petiole 2—20 cm long. **Inflorescence** of axillary, solitary flowers or short, few-flowered cymes. **Calyx** 3—5 mm long, shallowly 5-lobed, subtended by 2 or 3 linear bracts. **Corolla** of 5 obovate petals 3.5—8 mm long, lavender blue to pink, bilobed at the apex. Stamens numerous, monadelphous. Ovary superior. **Fruit** a rotate schizocarp 6—8 mm in diameter, breaking up into 10 ridged, triangular segments, surrounded by the enlarged, membranous, strongly veined calyx.

Malva parviflora L. is native from the Mediterranean to India, but is now widespread in temperate and tropical areas. It was first recorded from the Pacific Islands in 1826 (Hawai'i), where it is occasional in dry disturbed places at up to 2200 m elevation.

This herb can be distinguished by its alternate, round to kidney-shaped, 3—7-lobed leaves, foliage and stems covered with star-shaped hairs, flowers solitary or in short axillary cymes, corollas lavender blue to pink, numerous stamens fused into a column, and wheel-like fruits that split at maturity into 10 segments. Synonym: *Malva rotundifolia* of some Hawaiian authors.

MALVASTRUM COROMANDELIANUM
Malvaceae (Mallow Family)

COMMON NAMES: false mallow
DISTRIBUTION: all the main island groups except Belau (but rare in Samoa)

Widely branching subshrub up to 1 m in height. Stems appressed-pubescent with 4-branched hairs. **Leaves** alternate, simple, blade ovate to elliptic, 1.2—7 x 0.8—3.5 cm, acute at the apex, rounded to broadly cuneate at the base; surfaces appressed-pubescent; margins coarsely toothed; stipules linear-lanceolate; petiole 0.4—3 cm long. **Inflorescence** of 1—several flowers in short terminal and axillary clusters. **Calyx** 7—10 mm long, deeply lobed into 5 ovate, pubescent sepals, subtended by 3 narrow bracts 4—6 mm long. **Corolla** rotate, with 5 irregularly obovate, pale orange petals 7—12 mm long. Stamens many, monadelphous. Ovary superior. **Fruit** a rotate to subglobose schizocarp 4—7 mm in diameter, breaking into 10—15 one-seeded segments with a spine up to 2 mm long.

Malvastrum coromandelianum (L.) Garcke is native to tropical America, but is now widespread in the tropics. It was first recorded from the Pacific Islands in 1864 (Hawai'i), where it is occasional to common in disturbed places such roadsides, lawns, and waste places at up to 600 m elevation.

This subshrub can be distinguished by its widely branching stems, 4-parted hairs on the stems and foliage, alternate, distinctly veined leaves with coarsely toothed margins, 3 narrow bracts beneath the flowers, pale orange corollas, and wheel-like fruits that split into 10—15 segments. Synonym: *Malvastrum tricuspidatum* (R. Br.) A. Gray. Very similar to *Sida* species, which, however, lack the linear bracts beneath the flowers.

SIDA FALLAX
Malvaceae (Mallow Family)

COMMON NAMES: 'ilima (Hawai'i)
DISTRIBUTION: Hawai'i, Society Islands, Belau, Marshalls, Kiribati

Prostrate to erect shrub up to 1.5 m or more in height. Stems glabrous to densely tomentose, usually stellate. **Leaves** alternate, simple, blade lanceolate to suborbicular, mostly 1—8 x 1—7 cm, obtuse to acuminate at the apex, obtuse to subcordate at the base; surfaces mostly pubescent, lower densely so and duller; stipules linear, 3—9 mm long; petiole 0.5—5 cm long. **Inflorescence** of 1—7 flowers per axil, on pedicels 1—5 cm long. **Calyx** campanulate, 6—11 mm long, deeply divided into 5 triangular lobes. **Corolla** rotate, of 5 broadly obovate petals 8—20 mm long, yellow to orange, often red at the base, unevenly 2-lobed at the tip. Stamens many, monadelphous. Ovary superior, style lobes 5. **Fruit** a rotate schizocarp 4—6 mm in diameter, with 6—9 beaked or awned segments.

Sida fallax Walp. is native to the Pacific Islands, where it ranges from Hawai'i to China. In Hawai'i, where it is more abundant than elsewhere, it is occasional to common in relatively undisturbed places, particularly in coastal areas, where it is typically prostrate, and in mountain forests at up to 2000 m elevation, where it is erect. The flowers are commonly made into leis.

This shrub can be distinguished by its alternate leaves, dull green lower leaf surface, stalked, axillary flowers, showy orange bilobed petals, and wheel-like fruits of 6—9 segments with awns less than 1 mm long. Synonyms: *Sida cordifolia* of some Hawaiian authors *S. meyeniana* Walp. A related species in Hawai'i, Tonga (rare), Society Islands (rare), Guam, and Belau, *Sida cordifolia* L., has longer awns and 8—12 segments in the fruit.

SIDA RHOMBIFOLIA
Malvaceae (Mallow Family)

COMMON NAMES: broomweed, Cuba jute; mautofu (Samoa); te'ehosi (Tonga)
DISTRIBUTION: all the main island groups

Much-branched perennial shrub 30—150 cm in height. Stems finely stellate-pubescent. **Leaves** alternate, simple, blade rhomboid to ovate, 2—6.5 x 0.4—2.5 cm, acute to obtuse at the apex, cuneate to rounded at the base; lower surface densely stellate-pubescent; margins serrate; stipules linear; petiole 2—5 mm long. **Inflorescence** of solitary, axillary and subterminal flowers borne on a thin pedicel 1—3.5 cm long. **Calyx** cup-shaped, 4—7 mm long, deeply divided into 5 broadly ovate, apiculate, strongly ribbed lobes. **Corolla** rotate, 7—10 mm long, with 5 pale orange, obovate, unequally bilobed petals. Stamens numerous, monadelphous. Ovary superior. **Fruit** a flattened-globose schizocarp 4—7 mm in diameter, breaking up at maturity into 9—12 segments bearing a terminal spine

Sida rhombifolia L. is probably native to tropical America, but was an early introduction throughout the tropics. It was apparently an ancient introduction to the islands as far east as western Polynesia, and a European introduction to Hawai'i. It is common in disturbed areas such as waste places, plantations, and roadsides at up to 1200 m elevation. It is often a difficult weed to eradicate.

This shrub can be distinguished by its alternate, rhombic to ovate leaves, toothed leaf margins, solitary flowers lacking bracts beneath them, long thin, flower stalks, and wheel-like fruits splitting into about 10 sections, each bearing a terminal spine. A similar widespread species, *Sida acuta* Burm. f., differs in typically having the leaves in one plane and flowers on very short stalks.

SIDA SPINOSA
Malvaceae (Mallow Family)

COMMON NAMES: prickly sida
DISTRIBUTION: Hawai'i, Tahiti, Belau

Erect perennial subshrub 50—80 cm in height. Stems finely stellate-pubescent. **Leaves** alternate, simple, blade linear to oblong or ovate, 1—4 x 0.2—0.8 cm, acute at the apex, obtuse to truncate at the base; surfaces subglabrous to pubescent; margins serrate to crenate; stipules filiform, 2—5 mm long; petiole 2—12 mm long, often with 2 bumps on the adjacent stem. **Inflorescence** of solitary or clustered axillary flowers on a pedicel 2—10 mm long. **Calyx** cup-shaped, 4—7 mm long, strongly 10-ribbed. **Corolla** rotate, with 5 pale yellow to yellow-orange petals 5—7 mm long. Stamens numerous, monadelphous. Ovary superior. **Fruit** a subglobose schizocarp 4—6 mm in diameter, splitting into 5 puberulent segments 3—4 mm long, with 2 awns 0.5—1.5 mm long at the tip.

Sida spinosa L. is native to tropical America, but is now widely naturalized in warm regions of the world. It was first recorded in the Pacific Islands in 1971 (Hawai'i). It is particularly common in Hawai'i, where it is found in disturbed lowland areas, such as waste places and roadsides, especially on O'ahu, at up to 150 m or more in elevation.

This subshrub can be distinguished by its small, narrow, alternate leaves, toothed leaf margins, flowers solitary or in short clusters in the leaf axils, orange petals, subglobose, splitting fruits enclosed within the 10-nerved calyx, and fruit segments with 2 awns.

URENA LOBATA
Malvaceae (Mallow Family)

COMMON NAMES: hibiscus bur; mautofu (Samoa); mo'osipo (Tonga)
DISTRIBUTION: all the main island groups

Erect shrub up to 2 m in height. Stems often tinged purple, stellate-pubescent. **Leaves** alternate, simple, blade ovate to elliptic, 2—10 x 2—10 cm, acute at the apex, round to subcordate at the base; surfaces sparsely pubescent; margins irregularly toothed to deeply 3—7-lobed; stipules lanceolate, small; petiole 2—10 cm long. **Inflorescence** of solitary, axillary flowers on a pedicel 2—4 mm long. **Calyx** cup-shaped, 4—6 mm long, deeply divided into 5 ovate sepals subtended by an involucre of 5 bracts 4—7 mm long. **Corolla** rotate, 1.2—2 cm long, divided to near the base into 5 pink, obovate petals. Stamens numerous, monadelphous. Ovary superior. **Fruit** a schizocarp 8—12 mm across, splitting into 5 subglobose mericarps 2.5—3.5 mm long, covered with spines.

Urena lobata L. is probably native to tropical Asia and was an ancient introduction eastward to Micronesia and western Polynesia, and a European introduction to Hawai'i. It is occasional in disturbed places such as roadsides and pastures at up to 1300 m elevation, and perhaps is most common in Fiji, where it is a troublesome weed.

This shrub can be distinguished by its typically purple stems, alternate, ovate to deeply palmately lobed (on Guam and some other islands) leaves, short-stalked, axillary petals with pink flowers, stamens fused into a column, and 5-lobed fruits that split into 5 spiny segments. When lacking flowers, it is very similar to *Triumfetta rhomboidea* (see p. 134), which has much smaller yellow flowers and thinner, longer fruit spines.

CLIDEMIA HIRTA
Melastomataceae (Melastome Family)

COMMON NAMES: Koster's curse
DISTRIBUTION: Hawai'i, Samoa, Fiji, Belau

Coarse perennial shrub up to 2 m in height. Stems covered with red bristles that pale with age. **Leaves** opposite, simple, blade ovate to oblong, 5—18 x 3—10 cm, acuminate at the apex, rounded to subcordate at the base; surfaces pleated, 5-veined from the base, hispid; margins finely crenate; petiole 0.5—5 cm long, hispid. **Inflorescence** of short, hairy, axillary and terminal, 6—20-flowered panicles 1—3 cm long. **Calyx** of 5 hispid, linear lobes 2—4 mm long, atop the 3—4 mm long urceolate hypanthium. **Corolla** of 5 white, ovate petals 6—8 mm long. Stamens 10, with geniculate anthers. Ovary inferior. **Fruit** a hispid, ovoid, many-seeded, bluish-black berry 5—8 mm long, with the sepals persistent on top.

Clidemia hirta (L.) D. Don is native to tropical America and was first recorded in the Pacific Islands prior to 1905 (Fiji). It is common in relatively wet open or disturbed places, and is invasive and a serious pest in native forest areas at up to 1300 m elevation wherever it occurs. In some places it has been controlled by the introduction of a species of thrips (an insect).

This shrub can be distinguished by its hairy stems and foliage, opposite leaves with a pleated surface and 5 parallel veins from the base, short axillary cymes of white flowers, and small black, bristle-covered berries. Many other species of this family, Melastomataceae, are serious pests in Hawai'i.

FICUS MICROCARPA
Moraceae (Mulberry Family)

COMMON NAMES: Chinese banyan
DISTRIBUTION: Hawai'i, Guam, Belau, cultivated in Tahiti

Epiphyte or shrub that eventually grows into a large banyan tree with hanging aerial roots that may reach the ground and form pillars. Stems glabrous, with the tip enclosed within a sheath-like cap; sap milky. **Leaves** alternate, simple, blade oblong to obovate, 2.5—8 x 1.5—5 cm, rounded and often bluntly acuminate at the apex, round to acute at the base; surfaces glabrous, coriaceous, dark green; margins entire; petiole 0.4—2 cm long. **Inflorescence** an axillary syconium with the flowers lining the inside of a nearly closed globose structure, subtended by 3 broadly ovate bracts. **Fruit** the mature sessile syconium, subglobose, pink to red, 6—12 mm in diameter and containing numerous tiny seeds.

Ficus microcarpa L. f. is native from Sri Lanka to Micronesia, but is a European introduction to Hawai'i and Tahiti, where it was first recorded in the early 1900s (but it was not until 1938 that the wasp necessary for pollination was introduced to Hawai'i). It is occasional as an ornamental in Hawai'i and Tahiti, but is naturalized in disturbed lowland areas such as secondary forest, shrubland, and in cracks in walls and sidewalks. It occurs in limestone forest on Guam.

This banyan tree can be distinguished by its milky sap, terminal sheath on the stem, leathery, dark green leaves, and axillary, subglobose, pink to red fruit containing numerous tiny seeds. Synonym: *Ficus retusa* of many authors. Several other native species of banyans occur in the Pacific Islands (except Hawai'i).

PSIDIUM CATTLEIANUM
Myrtaceae (Myrtle Family)

COMMON NAMES: strawberry guava; waiawī (Hawai'i)
DISTRIBUTION: Hawai'i, Tahiti, Fiji, Belau, cultivated elsewhere

Shrub or small tree up to 6 m in height. Stems sparsely puberulent when young; bark flaky, smooth. **Leaves** opposite, simple, blade obovate to elliptic, 3—13 x 2—5 cm, bluntly cuspidate at the apex, attenuate to cuneate at the base; surfaces leathery, glabrous; margins entire; petiole 5—10 mm long. **Inflorescence** of solitary axillary flowers on a peduncle 4—8 mm long. **Calyx** of 4 rounded sepals 3—5 mm long, on the rim of the 6—8 mm long campanulate hypanthium. **Corolla** of 5 obovate, white, caducous petals 5—7 mm long. Stamens many, showy white. Ovary inferior. **Fruit** a globose to ellipsoid, red (rarely yellow) berry 2—3 cm in diameter, with a white pulp containing many smooth seeds.

Psidium cattleianum Sabine is native to tropical America and was first recorded from the Pacific Islands in 1825 (Hawai'i). It is common to abundant in disturbed and native, moderately wet to wet montane forest and scrubland in Hawai'i at up to 1200 m elevation, but reportedly grows in beach thickets in Fiji. The fruit is edible and tasty, but the tree is a harmful weed that takes over native forest areas.

This small tree can be distinguished by its smooth, flaky bark, leathery dark green opposite leaves usually with a finger-like extension at the tip, axillary solitary flowers, numerous showy white stamens, and red to yellow, ellipsoid to globose fruits with an edible pulp. Synonym: *Psidium littorale* Raddi. There are two yellow-fruited varieties that are uncommon in Hawai'i.

PSIDIUM GUAJAVA
Myrtaceae (Myrtle Family)

COMMON NAMES: guava; kuawa (Hawai'i); ku'ava (Samoa); kuava (Tonga)
DISTRIBUTION: all the main islands

Shrub or small tree up to 10 m in height. Stems finely pubescent and 4-angled when young, bark flaky, smooth. **Leaves** opposite, simple, blade elliptic to oblong, 6—15 x 3—7 cm, rounded to subacute at the apex, rounded to broadly cuneate at the base; upper surface glabrous, lower finely pubescent and strongly veined; petiole 3—8 mm long. **Inflorescence** of solitary flowers, or 1—3 per axil, on a peduncle 1.2—2.5 cm long. **Calyx** of 4 or 5 ovate sepals 7—10 mm long, on the rim of the 5—7 mm long campanulate hypanthium. **Corolla** of 5 white, elliptic, caducous petals 10—16 mm long. Stamens many, showy white. Ovary inferior. **Fruit** a globose to ovoid berry 5—10 cm long, yellow, with a pink or rarely yellow pulp containing many, compressed ovoid seeds 3—5 mm in diameter.

Psidium guajava L. is native to tropical America, but is now found throughout the tropics. It was first recorded from the Pacific Islands by the early 1800s (Hawai'i), where it is common in dry to wet, disturbed areas such as pastures, waste places, and scrub forest at up to 1200 m elevation. Originally introduced for its edible fruit, it is now a serious weed in pastures and elsewhere. The leaves are often boiled into a tea to treat diarrhea in many of the Pacific islands (Whistler 1992b).

This small tree can be distinguished by its smooth, flaky bark, opposite, elliptic to oblong leaves strongly veined on the lower surface, axillary flowers with numerous showy white stamens, and globose to ovoid fleshy fruits with a yellow rind and pink, seed-filled pulp.

SYZYGIUM CUMINI
Myrtaceae (Myrtle Family)

COMMON NAMES: Java plum
DISTRIBUTION: Hawai'i, Tahiti, Fiji, cultivated elsewhere

Large tree up to 20 m in height. Stems glabrous, bark pale yellow, flaky. **Leaves** opposite, simple, blade lanceolate to elliptic, 7—16 x 3—8 cm, bluntly acuminate to cuspidate at the apex, broadly cuneate to acute at the base; surfaces glabrous, leathery; margins entire; petiole 1—2.5 cm long. **Inflorescence** of compound cymes 4—9 cm long, axillary, terminal, or on old wood. **Calyx** of 4 tiny, caducous sepals *ca.* 0.5 mm long on the rim of the 3—5 mm long campanulate hypanthium. **Corolla** of 4 white petals united into a caducous cap 2.5—3.5 mm in diameter covering the flower. Stamens numerous, pink. Ovary inferior. **Fruit** an oblong to ellipsoid, asymmetrical berry 1.2—3 cm long, black or dark purple, glossy, 1-seeded.

Syzygium cumini (L.) Skeels is native from India to Malaysia, but is now widely cultivated in the tropics. It was first recorded from the Pacific Islands in 1871 (Hawai'i), where it is occasional to common in disturbed lowland places, such as lowland and valley forests, occasionally in scrubby vegetation in Hawai'i, and is perhaps also sparingly naturalized in Fiji. It is present in many other island groups only in cultivation.

This large tree can be distinguished by its leathery, opposite, lanceolate to elliptic leaves, short panicles of small flowers, numerous pink stamens (filaments), and shiny black to dark purple berries 1.2—3 cm long. Synonym: *Eugenia cumini* (L.) Druce. Numerous native species of this genus occur in the islands, but most of them are restricted to native habitats.

BOERHAVIA COCCINEA
Nyctaginaceae (Four-o'clock Family)

COMMON NAMES: none
DISTRIBUTION: Hawai'i, Guam

Prostrate to ascending perennial herb with a tuberous root. Stems sparingly glandular-puberulent. **Leaves** opposite, simple, blade broadly ovate to suborbicular, 1—4 x 1—3 cm, rounded to subacute at the apex, rounded to subcordate at the base; surfaces mostly glabrous; margins long-ciliate; petiole 0.5—2 cm long. **Inflorescence** a delicate, widely branching panicle up to 30 cm long, rising above the vegetative parts of the plant, bearing few-flowered clusters at the tips. **Calyx** campanulate, constricted near the middle, 1.5—2.5 mm long, magenta to dark red on the upper portion, glabrous, subsessile. **Corolla** absent. Stamens 2, free. Ovary superior. **Fruit** a clavate to obconical achene 3—4 mm long, enclosed within the calyx, with stalked glands along the 3—5 ribs.

Boerhavia coccinea Mill. is native to the Caribbean Islands and was first recorded from the Pacific Islands in 1974 (Hawai'i). It is common in disturbed places, such as roadsides and lawns, at up to 140 m elevation. It rapidly spreads by means of its sticky seeds, which are borne in diffuse inflorescences above the rest of the plant.

This low herb can be distinguished by its delicate, widely branching, erect inflorescences, tiny red flowers lacking a corolla, and tiny, sticky, cone- or club-shaped fruits. A similar widespread species, *Boerhavia repens* L., differs in having pink flowers clustered atop a long stalk. Several other native species are found in native lowland and coastal habitats (see Whistler 1992a).

LUDWIGIA OCTOVALVIS
Onagraceae (Evening-primrose Family)

COMMON NAMES: primrose willow; kāmole (Hawai'i)
DISTRIBUTION: all the main island groups

Erect perennial herb 60—200 cm in height. Stems sparingly to densely pubescent, longitudinally grooved, somewhat woody at the base. **Leaves** alternate, simple, blade linear to lanceolate, 5—12 x 0.7—2.5 cm, attenuate at the apex, cuneate to attenuate at the base; surfaces sparingly pubescent; margins entire; petiole 0—1 cm long. **Inflorescence** of solitary axillary flowers, borne on a pedicel 0.5—2 cm long. **Calyx** of 4 persistent, ovate to lanceolate sepals 8—12 mm long. **Corolla** of 4 caducous, yellow, obovate petals 10—16 mm long, notched at the tip. Stamens 8, free. Ovary inferior, 4-celled; stigma globose. **Fruit** a linear, longitudinally 8-ribbed capsule 2.5—5 cm long, containing numerous brown, compressed-ovoid or subglobose seeds 0.6—0.8 mm long.

Ludwigia octovalvis (Jacq.) Raven is native to tropical America, but is now pantropic in distribution. It was first recorded from the Pacific Islands as early as 1794 (Hawai'i), where it is now common to locally abundant in wetlands such as drainage ditches, pond margins, and taro patches at up to 1000 m elevation.

This herb can be distinguished by its alternate, narrow leaves, axillary, solitary flowers, 4 lanceolate sepals, 4 obovate yellow petals, and linear pods splitting open along the sides to release the tiny brown seeds. Synonyms: *Jussiaea erecta* of many authors, *J. suffruticosa* L. A similar species found in Samoa, Fiji, Guam, and Belau, *Ludwigia hyssopifolia* (G. Don) Exell, has smaller flowers with acute-tipped petals.

OXALIS CORNICULATA
Oxalidaceae (Wood-sorrel Family)

COMMON NAMES: wood sorrel; 'ihi'ai (Hawai'i); kihikihi (Tonga)
DISTRIBUTION: all the main island groups

Creeping perennial herb. Stems several from the main root, widely branching, pubescent, rooting from the nodes. **Leaves** alternate, palmately trifoliate, leaflet blades obcordate, 0.5—2 x 0.4—2.5 cm, rounded and deeply notched at the apex, broadly acute at the base; surfaces glabrous to villous; margins entire, ciliate; petiole 2—8 cm long. **Inflorescence** of 1—12 flowers in axillary, umbellate cymes. **Calyx** 2.5—4.5 mm long, divided to near the base into 5 lanceolate lobes. **Corolla** of 5 oblanceolate yellow petals 4—8 mm long. Stamens 10, in two series. Ovary superior, styles 5. **Fruit** a pubescent, many-seeded, 5-lobed, cylindrical to ellipsoid, acute-tipped capsule 9—19 mm long, opening explosively.

Oxalis corniculata L. is of uncertain origin, but is now cosmopolitan in distribution. It was an ancient introduction to the Pacific Islands, where it is occasional to common in disturbed places such as around houses, in lawns, and greenhouses, extending up to 2300 m elevation. It is used in traditional Polynesian remedies for ailments of infants (Whistler 1992b).

This creeping herb can be distinguished by its alternate, trifoliate leaves, heart-shaped leaflets, small yellow flowers in umbels, 10 stamens, and cylindrical pods that open explosively. Synonym: *Oxalis repens* Thunb. A similar species in Samoa, Tahiti, Fiji, and Belau, *Oxalis barrelieri* L., differs in being an erect herb with pink flowers. Another widespread species, *Oxalis corymbosa* DC., differs in being a stemless herb with larger pink flowers.

PASSIFLORA FOETIDA
Passifloraceae (Passionfruit Family)

COMMON NAMES: love-in-a-mist, wild passionfruit; pohāpohā (Hawai'i); pāsio vao (Samoa); vaine 'ae kumā (Tonga)
DISTRIBUTION: all the main island groups

Bad-smelling perennial vine. Stems up to 5 m long, densely hairy, with a coiled axillary tendril; vegetative parts often with gland-tipped hairs. **Leaves** alternate, simple, blade ovate, 3-lobed, 3—14 x 2—11 cm, acuminate at the apex, subcordate at the base; surfaces hispid; margins wavy; petiole 1.5—7 cm long. **Inflorescence** of solitary, axillary flowers; peduncle 3—6 cm long. **Calyx** of 5 oblong sepals 1.5—2.5 cm long, white on the inside, with a subterminal awn, entirely surrounded by pinnately branching bracts. **Corolla** of 5 white to pink petals similar to the sepals but awnless. Stamens 5, surrounded by a corona of filaments all white, or purple at the base. Ovary superior, 3-lobed; stigmas 3. **Fruit** a thin-walled, globose, red to orange berry 1.5—2.5 cm in diameter.

Passiflora foetida L. is native to tropical America, but is now a pantropic weed. It was first recorded from the Pacific Islands in 1871 (Hawai'i), where it is common to locally abundant growing over low vegetation in disturbed places, such as pastures and roadsides at up to 500 m elevation, and also on lava flows. The small orange fruits are edible.

This vine can be distinguished by its hairy foliage, alternate, 3-lobed leaves, densely coiled, axillary tendrils, showy white to lavender flowers with an all white to white and purple corona, and red to orange globose fruits surrounded by the pinnately branching bracts. The naturalized edible passionfruit, *Passiflora edulis* Sims, is mostly glabrous, has large, edible, yellow fruits, and lacks the pinnately branching flower bracts.

PASSIFLORA SUBEROSA
Passifloraceae (Passionfruit Family)

COMMON NAMES: wild passionfruit; huehue haole (Hawai'i)
DISTRIBUTION: Hawai'i, Samoa (rare), Tahiti, Fiji, Guam

Perennial herbaceous vine. Stems striate, softly pubescent to glabrous with a coiled tendril borne in the axil. **Leaves** alternate, simple, ovate to deeply 3-lobed, 3—11 x 1.5—9 cm, acute at the apex, rounded to subcordate at the base; surfaces subglabrous, or pubescent on the veins; margins entire, ciliate or glabrous; stipules linear-lanceolate, 3—6 mm long; petiole 0.5—2 cm long, with a conspicuous pair of glands near the middle. **Inflorescence** of paired axillary peduncles 0.5—2 cm long, bearing 1 or occasionally 2 flowers. **Calyx** of 5 ovate, greenish white sepals 6—12 mm long. **Corolla** absent. Stamens 5, surrounded by a corona of filaments white or white with purple at the base. Ovary superior. **Fruit** a globose to ovoid, dark purple berry 6—18 mm long.

Passiflora suberosa L. is native to tropical America, but is now widely naturalized in the tropics. It was first recorded from the Pacific Islands in 1916 (Hawai'i), where it is occasional to common in disturbed places, such as shrublands in gardens, but is also naturalized in open native forest in Hawai'i and perhaps elsewhere, at up to 600 m elevation. Rare in Samoa.

This vine can be distinguished by its often lobed leaves, coiled axillary tendrils, small flowers with greenish white petaloid sepals but no petals, white or purple corona, and small, dark purple, globose to ovoid fruits. It is quite different from *Passiflora foetida*, which is much hairier and has the orange to red fruit enclosed within pinnately branching bracts.

RIVINA HUMILIS
Phytolaccaceae (Pokeweed Family)

COMMON NAMES: coral berry; polo (Tonga)
DISTRIBUTION: Hawai'i, Tonga, Tahiti, Fiji

Erect slender herb 40—100 cm in height. Stems glabrous, longitudinally grooved. **Leaves** alternate, simple, blade lanceolate, 4—10 x 1.5—4 cm, acute to acuminate at the apex, acute to rounded at the base; surfaces glabrous or puberulent (especially along the veins); margins entire; petiole 1—3.5 cm long. **Inflorescence** a many-flowered axillary raceme 4—12 cm long. **Calyx** 2—3.5 mm long, deeply divided into 4 oblanceolate sepals, white or pink, subtended by 2 minute bracteoles and 1 bract; pedicel 2—4 mm long. **Corolla** absent. Stamens 4, free. Ovary superior. **Fruit** a red, or sometimes orange or purple, subglobose drupe 3—4 mm in diameter, containing a single seed.

Rivina humilis L. is native to subtropical and tropical America and was first recorded from the Pacific Islands in 1871 (Hawai'i). It is occasional in disturbed, shady or sunny, lowland places such as scrub forest, roadsides, and pastures.

This herb can be distinguished by its alternate leaves, axillary and terminal racemes of small flowers, white to pink calyces of 4 sepals, absence of a corolla, and small red, orange, or purple drupes. The fruit is superficially similar to *Solanum* spp., but the presence of racemes rather than short umbels or clusters serves to distinguish it from these species.

PLANTAGO MAJOR
Plantaginaceae (Plantain Family)

COMMON NAMES: broad-leaved plantain, common plantain; laukahi (Hawai'i)
DISTRIBUTION: all the main island groups except Belau

Stemless perennial herb with fibrous roots. **Leaves** in a basal rosette, blade ovate to broadly elliptic, mostly 3—15 x 2—9 cm, obtuse at the apex, abruptly cuneate at the base; surfaces glabrous to pubescent, 5-nerved; margins usually with a few tiny teeth; petiole mostly 2—12 cm long, winged. **Inflorescence** a narrow cylindrical spike 10—25 or more cm long, with the flowers crowded on the upper half. **Calyx** of 4 ovate to suborbicular brown, membranous sepals 1.5—2.5 mm long, subtended by a membranous, ovate to triangular bract 1—2 mm long. **Corolla** sympetalous, membranous, brown, 2—3 mm long, divided at the top into 4 spreading, lanceolate lobes. Stamens 4, free, long-exserted. Ovary superior. **Fruit** a brown, ellipsoid, several-seeded, circumscissile capsule 2—3 mm long.

Plantago major L. is native to Europe and temperate Asia, but is now a cosmopolitan weed. It was first recorded from the Pacific Islands in 1864 (Hawai'i), where it is occasional to locally common in sunny, disturbed places, such as lawns and pastures at up to 1200 m elevation, as well as in disturbed rain forest in Hawai'i. The leaves are used in traditional Hawaiian remedies (Whistler 1992b), and the seeds have laxative properties.

This stemless herb can be distinguished by its basal, parallel-veined, broadly ovate leaves, loose spikes, membranous flower parts, and exserted stamens. A similar widespread species, *Plantago lanceolata* L., has narrower leaves and dense ovoid spikes. In Hawai'i and Tonga, *Plantago debilis* R. Br., has loose spikes and shorter, oblanceolate to obovate leaves less than 2 cm wide.

POLYGALA PANICULATA
Polygalaceae (Milkwort Family)

COMMON NAMES: bubble-gum plant; pulunamulole (Samoa)
DISTRIBUTION: all the main island groups except Tonga

Annual, erect to ascending herb up to 75 cm in height with roots redolent of bubble gum. Stems much branched, puberulent. **Leaves** alternate or the lower ones often whorled, simple, blade oblong to linear, 8—32 x 1—5 mm, acute at the apex, attenuate at the base, subsessile; surfaces glabrous; margins entire; petiole 0—2 mm long. **Inflorescence** in axillary or pseudoterminal slender racemes 3—15 cm long. **Calyx** of 5 white, unequal sepals, outer 3 ovate, 0.7—1.2 mm long, inner pair petaloid, elliptic to oblanceolate, 1.8—2.5 mm long. **Corolla** of 3 white petals similar to inner sepals. Stamens 8, fused into a sheath split along one side. Ovary superior. **Fruit** an oblong capsule, 2—2.5 mm long, containing two black, pubescent, ellipsoid seeds.

Polygala paniculata L. is native to tropical America and was first recorded from the Pacific Islands in 1905 (Samoa). It is occasional to common, but never dominant, in disturbed places such as roadsides, cane fields, gardens, and plantations at up to 1100 m elevation, and is naturalized on lava flows in Samoa.

This small, delicate herb can be distinguished by its branching stems, small, narrow, alternate leaves, tiny white flowers in narrow axillary and subterminal racemes, and especially by its root, which, when freshly pulled up, smells like bubble gum.

PORTULACA OLERACEA
Portulacaceae (Purslane Family)

COMMON NAMES: purslane, pigweed; ʻākulikuli kula (Hawaiʻi); tamole (Samoa, Tonga)
DISTRIBUTION: all the main island groups

Prostrate to ascending herb up to 30 cm in height. Stems glabrous except at the axils, succulent, branching at the base, often reddish. **Leaves** alternate to subopposite, simple, blade obovate to cuneate, 0.5—2.5 x 0.3—0.8 cm, obtuse to slightly notched at the apex, cuneate at the base, subsessile; surfaces glabrous; margins entire. **Inflorescence** of cymose clusters of 2—6 sessile flowers at the branch tips. **Calyx** of 2 green, ovate sepals 2—4 mm long. **Corolla** of 5 obovate petals 3—10 mm long, notched at the tip, yellow. Stamens 7—12, free. Ovary superior, 1-celled; style 5-branched. **Fruit** an ovoid, circumscissile capsule 4—5 mm long, opening by a long terminal cap that splits off to release the numerous, lens-shaped, black seeds 0.5—0.7 mm long.

Portulaca oleracea L. is of uncertain origin, but is now cosmopolitan in distribution. It was first recorded from the Pacific Islands in 1840 (Tonga), where it is common in disturbed places, such as gardens, plantations, and roadsides, at up to 900 m elevation, and is also naturalized in relatively undisturbed coastal areas.

This prostrate succulent herb can be distinguished by its fleshy, often reddish stems, mostly obovate fleshy leaves, terminal clusters of several yellow flowers, and capsules with a cap that splits off to release the tiny black seeds. Several native species are also found in littoral areas in the region (see Whistler 1992a).

PORTULACA PILOSA
Portulacaceae (Purslane Family)

COMMON NAMES: none
DISTRIBUTION: Hawai'i

Prostrate, mat-forming perennial herb with a thick taproot. Stems up to 20 cm long, purple to red, glabrous but with a dense tuft of hairs 2—6 mm or more long at the axils. **Leaves** alternate or clustered at the ends of the stems, simple, succulent, blade narrowly oblong to linear, 4—13 x 0.5—3 mm, acute at the apex, acuminate to cuneate at the base; surfaces glabrous; margins entire; petiole 1—2 mm long. **Inflorescence** of 2—8 flowers in a dense terminal cyme, subtended by a cluster of leaves, small bracteoles, and a dense tuft of hairs. **Calyx** of 2 triangular sepals 2.5—4.5 mm long. **Corolla** of 5 obovate, magenta petals 5—7 mm long. Stamens mostly 20—30, free, yellow. Ovary superior. **Fruit** an ellipsoid capsule 2—4 mm long, with the top splitting off to release the many tiny black seeds.

Portulaca pilosa L. is of uncertain origin and is pantropic in distribution. It was first recorded from the Pacific Islands in 1922 (Hawai'i), where it is common as a weed of dry, disturbed lowland places, such as open shrubland, often on rocks and on lava flows.

This succulent herb can be distinguished by its prostrate habit, narrow, succulent leaves, dense tufts of long hairs in the axils, terminal clusters of magenta flowers with 5 petals, and capsules with the top splitting off to release the numerous tiny black seeds. Synonym: *Portulaca cyanosperma* Egler. Other species of *Portulaca* in various Pacific archipelagoes have been incorrectly identified as this species, but these all have yellow flowers.

GREVILLEA ROBUSTA
Proteaceae (Protea Family)

COMMON NAMES: silk oak, silver oak; ‘oka kilika (Hawai’i)
DISTRIBUTION: Hawai’i, cultivated elsewhere

Tall tree up to 25 m or more in height. Stems rusty-tomentose when young. **Leaves** alternate, compound, fern-like, bipinnatifid, mostly 10—30 cm long, pinnately divided into 10—20 lanceolate to linear, irregularly lobed leaflets 4—12 cm long, acuminate at the apex, acuminate and oblique at the base; upper surface mostly glabrous, lower white-tomentose; margins of leaflets mostly deeply 2—6 lobed. **Inflorescence** in subterminal, one-sided racemes or a panicle of racemes, 7—15 cm long. **Calyx** salverform, curved, orange to golden brown, 6—9 mm long, 4-lobed, splitting along one side. **Corolla** absent. Stamens 4. Ovary superior, style protruding through the split in the calyx, persistent. **Fruit** an oblique, boat-shaped follicle 1.5—2 cm long, splitting to release the many winged seeds.

Grevillea robusta A. Cunn. ex R. Br. is native to Australia, but is now widespread in cultivation. It was first recorded from the Pacific Islands in 1880 (Hawai’i), where it is occasional in disturbed lowland areas, such as scrub and secondary forest, and sometimes on disturbed lava flows. It is cultivated on many islands in the tropical Pacific, but is noted to be naturalized only in Hawai’i.

This large tree can be distinguished by its alternate, fern-like leaves with silky white pubescence on the lower surface, one-sided panicles, orange petal-like sepals in a curved corolla tube that splits down one side, and boat-shaped fruits that split along one side to release winged seeds. Synonym: *Stylurus robusta* (A. Cunn. ex R. Br.) Degener.

RUBUS ROSIFOLIUS
Rosaceae (Rose Family)

COMMON NAMES: thimbleberry; ōla'a (Hawai'i)
DISTRIBUTION: Hawai'i, Tahiti

Erect to trailing shrub up to 2 m or more in height. Stems pilose, sparsely covered with prickles 1—4 mm long. **Leaves** alternate, pinnately compound, 7—18 cm long, with 3—7 leaflets, leaflet blades ovate to lanceolate, 1.5—8 x 0.8—3.5 cm, acuminate to acute at the apex, rounded to truncate at the base; surfaces pilose to nearly glabrous; margins doubly serrate; stipules filiform, 6—10 mm long. **Inflorescence** mostly of solitary, terminal or axillary flowers on a pedicel 0.5—3 cm long. **Calyx** of 5 lanceolate sepals 1.4—2.5 cm long, tomentose, apex long-attenuate. **Corolla** of 5 white, obovate petals 1—2 cm long. Stamens many, free. Ovaries many. **Fruit** a subglobose, red, multiple fruit 2—3.5 cm long, easily detaching from the receptacle.

Rubus rosifolius Sm. is native to Asia, but is now widespread in the tropics. It was first recorded from the Pacific Islands in the 1880s (Hawai'i), where it is occasional to common in moist, disturbed places such as along roadsides and in disturbed forest in Hawai'i and Tahiti, ranging from 60 to 1700 m elevation. The fruit is edible and tasty.

This prickly shrub can be distinguished by its alternate, pinnately compound leaves, solitary axillary flowers, 5 showy white, obovate petals, many stamens, many ovaries, and red multiple fruits that are easily separated from the receptacle. Another member of the same genus, *Rubus argutus* Link, is a noxious weed in Hawai'i, and differs in having a scrambling habit, leaves with 3 or 5 leaflets, and black berries.

SPERMACOCE ASSURGENS
Rubiaceae (Coffee Family)

COMMON NAMES: buttonweed
DISTRIBUTION: all the main island groups

Erect annual or short-lived perennial herb up to 50 cm in height. Stems glabrous to finely pubescent, round or somewhat 4-angled. **Leaves** opposite, simple, blade narrowly elliptic, 2—6 x 0.6—1.4 cm, acute to attenuate at the apex, cuneate to attenuate at the base; surfaces glabrous, often tinged purple; margins entire; stipules interpetiolar, 1—2.5 mm long with a ridge of bristle up to twice as long; petiole 1—4 mm long. **Inflorescence** of dense, sessile, axillary and terminal clusters within the stipules. **Calyx** of 4 narrow sepals 0.5—0.7 mm long atop the 2—3 mm long hypanthium. **Corolla** funnelform, 2—3 mm long, white, with 4 triangular, spreading lobes. Stamens 4, epipetalous, included. Ovary inferior. **Fruit** an ellipsoid capsule 2—3 mm long, containing 2 brown, narrowly oblong seeds 1.5—2.2 mm long, with a groove on one side.

Spermacoce assurgens Ruiz and Pavon is native to tropical America, and was first recorded from the Pacific Islands in 1929 (Hawai'i). It is occasional to common in disturbed places such as roadsides, lawns, and plantations at up to 850 m elevation.

This wiry herb can be distinguished by its opposite, narrowly elliptic leaves often tinged purple, interpetiolar stipules with a ridge of bristles, dense axillary clusters of flowers between the stipules, and small, white, 4-lobed corolla. Synonym: *Borreria laevis* of many authors, not (Lam.) Griseb. A similar species in Hawai'i, Samoa, and Fiji, *Spermacoce mauritiana* Gideon, differs in having smaller fruits and flowers. Another similar species in Samoa, Fiji, and Belau, *Spermacoce bartlingiana* (DC.) Fosb., differs in having broader leaves and winged, hairy stems.

TRIPHASIA TRIFOLIA
Rutaceae (Citrus Family)

COMMON NAMES: limeberry; limon-china (Guam)
DISTRIBUTION: Samoa (rare), Fiji, Guam

Small tree up to 4 m in height. Stems somewhat zigzag, puberulent, with paired axillary spines 3—15 mm long. **Leaves** alternate, trifoliate, leaflet blades ovate to suborbicular, terminal one 2.5—5 x 1.2—3 cm (lateral ones smaller), notched and bluntly obtuse to obtuse at the apex, cuneate to subround at the base; surfaces glabrous, gland-dotted; margins shallowly crenate; petiole 2—4 mm long. **Inflorescence** usually of solitary axillary flowers on a pedicel 1—2 mm long. **Calyx** 1.5—2.5 mm long, divided about halfway into 4 lobes, pubescent. **Corolla** of 3 oblanceolate, white petals 1—1.4 cm long. Stamens 6, free. Ovary superior. **Fruit** a red, subglobose to ellipsoid berry 9—13 mm long, glandular-punctate.

*Triphasia trifolia (*Burm. f.) P. Wils. is native to Southeast Asia and Malaysia, but was probably an ancient introduction to the islands of western Micronesia, and a recent introduction to the southern U.S. and Western Samoa (where it is rare). It is common to abundant in disturbed and native scrub forest on limestone areas in Guam, sometimes forming dense, spiny thickets. The fruits are edible but somewhat acrid. Its thorns make it a pest, especially when it is dense enough to form thickets.

This shrub may be distinguished by its paired axillary spines, alternate, trifoliate leaves, showy white flowers, 3 petals, 6 stamens, and fleshy red berries 9—13 mm long and gland-dotted on the surface.

NICANDRA PHYSALODES
Solanaceae (Nightshade Family)

COMMON NAMES: apple of Peru
DISTRIBUTION: Hawai'i

Erect annual herb up to 1.5 m in height. Stems mostly glabrous, angled, hollow. **Leaves** alternate, simple, blade ovate, 4—20 x 2—10 cm, acute to acuminate at the apex, cuneate to attenuate and often oblique at the base; surfaces mostly glabrous; margins irregularly dentate to subentire; petiole 1—6 cm long. **Inflorescence** of solitary, axillary flowers. **Calyx** 1.5—2.2 cm long, divided to near the base into 5 ovate lobes sagittate at the base and mucronate at the tip. **Corolla** sympetalous, broadly campanulate, 2.5—3.2 cm long, pale blue to mauve with a blue spot at the base of each lobe, shallowly 5-lobed. Stamens 5, epipetalous. Ovary superior. **Fruit** a dry, pale yellow, globose berry 1—2 cm in diameter, enclosed within the scarious calyx 2.5—3.2 cm long.

*Nicandra physalodes (*L.) Gaertn. is native to Peru and was first recorded from the Pacific Islands in 1871 (Hawai'i). It is occasional in disturbed places in Hawai'i, such as roadsides and waste places, at up to 830 m elevation. It was originally cultivated as an ornamental for its attractive flowers (Wagner *et al.* 1990).

This robust herb can be distinguished by its alternate leaves with entire to toothed margins, solitary axillary flowers, showy pale blue corollas with darker spots at the base, and bladder-like calyx surrounding the fruit. It is similar to *Physalis* spp., but differs in having larger leaves, larger blue rather than yellow flowers, and calyx lobes that are sagittate at the base.

NICOTIANA GLAUCA
Solanaceae (Nightshade Family)

COMMON NAMES: Indian tobacco, tree tobacco; mākahāla (Hawai'i)
DISTRIBUTION: Hawai'i

Shrub or small tree up to 5 m or more in height. Stems glabrous. **Leaves** alternate, simple, blade ovate, mostly 4—10 x 2—8 cm, acute to obtuse at the apex and base; surfaces glabrous, glaucous, gray-green; margins entire; petioles 2—6 cm long. **Inflorescence** of dense, many-flowered, terminal and upper axillary panicles 8—25 cm long, bearing small lanceolate bracts 2—6 mm long. **Calyx** tubular, 8—15 mm long, shallowly divided into 5 triangular lobes; pedicel 5—15 mm long. **Corolla** yellow, tubular, mostly 3—4 cm long, shallowly divided into 5 lobes, pubescent on the outside. Stamens 5, epipetalous. Ovary superior. **Fruit** an ovoid to subglobose capsule 8—13 mm long, enclosed within the persistent calyx, containing numerous tiny black seeds.

Nicotiana glauca R. C. Graham is native to Argentina, but is now widely naturalized in the warm-temperate areas of the world. It was first recorded from the Pacific Islands in 1864 (Hawai'i), where it is locally common in dry, disturbed places in the lowlands of Hawai'i (O'ahu, Lāna'i, Maui, and Kaho'olawe) at up to 350 m elevation. It was originally introduced as an ornamental plant, but is hardly used for this today.

This shrub may be recognized by its large, alternate, glaucous, gray-green leaves, flowers in terminal and upper axillary panicles, yellow tubular corollas, and ovoid to subglobose capsules containing numerous tiny black seeds. The cultivated and naturalized tobacco plant, *Nicotiana tabacum* L., differs in having glandular-pubescent leaves and white to red flowers.

PHYSALIS ANGULATA
Solanaceae (Nightshade Family)

COMMON NAMES: wild cape-gooseberry; vīvao (Samoa); polopā (Tonga)
DISTRIBUTION: all the main island groups

Erect, much-branched annual herb 20—100 cm in height. Stems angled, hollow, mostly glabrous. **Leaves** alternate, simple, blade ovate to elliptic, 2—12 x 1—5 cm, acute at the apex, cuneate to rounded and oblique at the base; surfaces mostly glabrous; margins entire to irregularly toothed; petiole 0.5—5.5 cm long. **Inflorescence** of solitary, axillary, nodding flowers. **Calyx** ovoid, 2—5 mm long, divided less than halfway into 5 triangular lobes; pedicel 0.6—1.5 cm long. **Corolla** sympetalous, rotate, 6—9 mm long, shallowly 5-lobed, pale yellow with darker spots in the center. Stamens 5, epipetalous, pale blue. Ovary superior. **Fruit** a subglobose berry 1—1.5 cm long, surrounded by the membranous, inflated, urn-shaped calyx 2.2—3.2 cm long.

Physalis angulata L. is native to tropical America and may be native to the Pacific Islands as well, since it was first recorded there (Tahiti) in 1769. It is occasional to common in disturbed lowland places, such as fallow fields and roadsides, but is not common in Hawai'i. The small fruits enclosed within the bladder-like calyx are edible.

This herb can be distinguished by its alternate leaves, solitary, axillary, wheel-like flowers that are pale yellow with brown at the base, and small globose berries surrounded by a bladder-like calyx. Synonym: *Physalis minima* L. A related species that is more common in Hawai'i, but less common elsewhere, is *Physalis peruviana* L., which differs in being pubescent and having a larger corolla. Its fruits are also edible.

SOLANUM AMERICANUM
Solanaceae (Nightshade Family)

COMMON NAMES: black nightshade; pōpolo (Hawai'i); polo, māgalo (Samoa); polo kai (Tonga); boro (Fiji)
DISTRIBUTION: all the main island groups

Erect, much-branching annual herb up 1 m in height. Stems somewhat angled, glabrous to slightly pubescent. **Leaves** alternate, simple, blade ovate to broadly lanceolate, 2—9 x 1—5 cm, acute to acuminate at the apex, cuneate to attenuate and winged at the base; surfaces mostly glabrous; margins entire to shallowly and irregularly lobed; petioles 0.5—3 cm long. **Inflorescence** a few-flowered umbel-like cyme 1.5—3 cm long, borne on the stems. **Calyx** campanulate, 1.5—2.5 mm long, shallowly divided into 5 obtuse, reflexed lobes; pedicel 4—10 mm long. **Corolla** stellate, 2.5—5 mm long, white, deeply divided into 5 spreading, lanceolate lobes. Stamens 5, epipetalous, yellow, exserted. Ovary superior. **Fruit** a shiny black, many-seeded, globose berry 4—6 mm long.

Solanum americanum Mill. is apparently a native species in the Pacific Islands, although it probably originated somewhere in tropical America. It is occasional in disturbed areas and sunny places in native forest at up to 2300 m elevation. The fruits are edible, and the leaves have commonly been used for external medicines in Polynesia (Whistler 1992b).

This herb can be distinguished by its small alternate leaves, umbel-like cymes of flowers borne on the stems, white, star-shaped corollas, yellow stamens, and small, glossy black berries. Synonyms: *Solanum nigrum* of many authors, not L., *S. nodiflorum* Jacq., *S. oleraceum* of some authors. Several species of *Solanum* are found in Polynesia, some of them, such as *Solanum torvum* (see P. 132), have nasty spines.

SOLANUM LINNAEANUM
Solanaceae (Nightshade Family)

COMMON NAMES: apple of Sodom, Sodom's apple; pōpolo kīkānia (Hawai'i)
DISTRIBUTION: Hawai'i, Fiji

Spreading shrub up to 1.8 m in height. Stems stellate-pubescent, armed with straight yellow prickles up to 15 mm long. **Leaves** alternate, simple, blade ovate to elliptic, 3—12 x 4—7 cm, acute to rounded at the apex, truncate to subcordate and oblique at the base; surfaces pubescent, midrib prickly; margins usually deeply cut into 5—7 irregular lobes; petioles 1—2 cm long. **Inflorescence** an axillary racemose cyme 1—5 cm long, with up to 6 flowers. **Calyx** campanulate, 6—8 mm long, shallowly 5-lobed, pedicel 6—12 mm long. **Corolla** rotate, 1.8—2.5 cm in diameter, shallowly divided into 5 acute-tipped lobes, white. Stamens 5, yellow. Ovary superior. **Fruit** a yellow, many-seeded, globose berry 2—3 cm in diameter, drying brown and firm, stalk prickly.

Solanum linnaeanum Hepper & P. Jaeger is native to either southern Europe or South Africa and was introduced to the other at an early date. It was first recorded from the Pacific Islands in 1895 (Hawai'i), where it is occasional in disturbed areas at up to 575 m elevation, such as roadsides, pastures, and shrublands in Hawai'i, but is less common in Fiji. Because of its thorns, it is a serious pest, but is not usually abundant where it does occur.

This herb can be distinguished by its prickly stems, alternate, ovate to elliptic leaves, pubescence of star-shaped hairs, white, wheel-shaped flowers, yellow stamens, and yellow, globose berries. Synonym: *Solanum sodomeum* of some authors, not L.

SOLANUM SEAFORTHIANUM
Solanaceae (Nightshade Family)

COMMON NAMES: none
DISTRIBUTION: Hawai'i

Climbing perennial vine. Stems up to 4 m in length, thin, mostly glabrous. **Leaves** alternate, simple, pinnatifid, blade ovate in outline, 5—11 x 4—8 cm, acute to acuminate at the apex, oblique at the base; margins divided to the midrib into 3—9 irregular lobes; surfaces glabrous; petiole 2—4 cm long. **Inflorescence** of many-flowered terminal and axillary panicles. **Calyx** tubular, 1—2 mm long, shallowly divided into 5 rounded lobes; pedicel 6—12 mm long. **Corolla** star-shaped, 2—3 cm in diameter, deeply divided into 5 lanceolate lobes, mauve blue. Stamens 5, epipetalous, yellow. Ovary superior. **Fruit** a globose, succulent berry *ca.* 1 cm in diameter, bright red.

Solanum seaforthianum Andr. is native to the West Indies and was first recorded from the Pacific Islands in 1916 (Hawai'i). It is occasional climbing in trees in *Leucaena* thickets and other disturbed lowland areas of Hawai'i. It was probably originally introduced to Hawai'i as an ornamental, or for its fruits that are used in making leis, but is now naturalized.

This vine can be distinguished by its lack of tendrils, alternate, deeply lobed leaves, panicles of showy bluish flowers, and red, globose, tomato-like fruits. It differs from all the other local species of the genus *Solanum* in having divided leaves and in being a vine.

SOLANUM TORVUM
Solanaceae (Nightshade Family)

COMMON NAMES: prickly solanum; tisaipale (Tonga)
DISTRIBUTION: all the main island groups

Erect, branching perennial shrub up to 4 m in height. Stems densely stellate-tomentose, with scattered prickles 2—9 mm long. **Leaves** alternate, simple, blade ovate to elliptic, mostly 5—20 x 5—15 cm, acute to obtuse at the apex, rounded to sagittate and often oblique at the base; surfaces stellate-pubescent, lower surface densely so; margins irregularly lobed; petioles 2—5 cm long. **Inflorescence** a many-flowered corymb 3—10 cm long, borne at intervals on the stem. **Calyx** campanulate, 3—10 mm long, divided over halfway into 5 acute lobes; pedicel 5—15 mm long. **Corolla** rotate, 12—18 mm long, white, deeply divided into 5 lanceolate lobes. Stamens 5, epipetalous, yellow. Ovary superior. **Fruit** a green to yellow, many-seeded, globose berry 1—1.5 cm long.

Solanum torvum Sw. is native to the Caribbean and was first recorded from the Pacific Islands in 1906 (Fiji). It is locally common to abundant in disturbed places such as fallow land, often forming dense thickets at up to 900 m elevation. In Hawai'i it is apparently limited so far to O'ahu and Maui, but can be expected to spread to the other islands.

This shrub can be distinguished by its prickly and tomentose vegetative parts, large, irregularly lobed or toothed, alternate leaves, white flowers in cymes borne above the axils, yellow stamens, and green or yellowish globose fruits. It is very similar to *Solanum mauritianum* Scop. of Hawai'i, Tonga, and Fiji, which differs in lacking prickles and having small axillary leaves.

WALTHERIA INDICA
Sterculiaceae (Cacao Family)

COMMON NAMES: ‘uhaloa (Hawai’i)
DISTRIBUTION: all the main island groups except Belau

Scarcely branched, erect to ascending shrub 1.5 m in height. Stems and other parts tomentose with stellate hairs. **Leaves** alternate, simple, blade ovate to elliptic, 2—10 x 1—6 cm, rounded at the apex, rounded to subacute at the base; surfaces tomentose, rugose; margins serrate; petiole 0.5—4 cm long. **Inflorescence** of short, dense, subglobose to irregularly ovoid axillary clusters 0.6—3 cm long, borne on a peduncle 2—12 mm long. **Calyx** campanulate, 3—5 mm long, divided about halfway into 5 lanceolate lobes. **Corolla** of 5 spathulate petals 3.5—5 mm long, yellow drying to orange. Stamens 5, free. Ovary superior. **Fruit** an obovoid, 2-valved capsule 2.5—3 mm long, enclosed within the persistent calyx, containing 1 black, top-shaped seed.

Waltheria indica L. is native to tropical America, and possibly to the Pacific Islands, at least to Hawai’i, where it was recorded during Captain Cook’s visit there in 1779, and is now pantropical in distribution. It is common in dry, disturbed areas, such as scrub forest, roadsides, and waste places, but is now rare in Tonga and Samoa and most of the other Polynesian islands. The roots have long been used in traditional Hawaiian remedies to treat sore throat (Whistler 1992b).

This shrub can be distinguished by its stems and other vegetative parts being covered with a thick, velvety layer of star-shaped hairs, alternate, ovate to elliptic leaves with a rugose surface, hairy, dense axillary clusters of flowers, small yellow corollas, and small capsules containing a single black, top-shaped seed. Synonym: *Waltheria americana* L.

TRIUMFETTA RHOMBOIDEA
Tiliaceae (Linden Family)

COMMON NAMES: bur bush; mautofu (Samoa); mo'osipo (Tonga)
DISTRIBUTION: all the main island groups

Erect shrub up to 1.5 m in height. Stems sparsely or densely pubescent with stellate hairs. **Leaves** alternate, simple, blade ovate to rhombic, sometimes 3—5-lobed, 2—10 x 0.8—9 cm, acuminate at the apex, cuneate to truncate at the base; surfaces densely stellate-pubescent, especially the lower surface, palmately veined; margins irregularly serrate; petiole 0.5—6 cm long. **Inflorescence** of short, dense, axillary and terminal cymes mostly less than 1.5 cm in diameter. **Calyx** deeply divided into 5 pubescent, narrowly oblong sepals 4—5 mm long. **Corolla** deeply divided into 5 yellow, obovate petals 3.5—5 mm long. Stamens 10—15, free. Ovary superior. **Fruit** a subglobose bur with a pubescent body 3—4 mm in diameter, covered with 75—100 hooked, glabrous spines 1—1.5 mm long.

Triumfetta rhomboidea Jacq. is native to tropical America and was first recorded from the Pacific Islands in 1891 (Tonga). It is occasional to locally common in disturbed places, such as plantations, roadsides, and pastures, at up to 760 m elevation. The burs, which are covered with hooked bristles, allow it to easily spread by sticking to fur or clothing, making it a pest.

This shrub can be distinguished by its dense, star-shaped hairs, alternate, ovate to rhombic or 3—5-lobed leaves, small yellow flowers with 10—15 stamens, and pubescent burs covered with glabrous, hooked bristles. Synonym: *Triumfetta bartramia* L. A very similar species of Hawai'i, Samoa (rare there), and Guam, *Triumfetta semitriloba* Jacq., differs in having a glabrous bur with bristles on the hooked spines.

LAPORTEA INTERRUPTA
Urticaceae (Nettle Family)

COMMON NAMES: island nettle; ogoogo toto (Samoa); hogohogo (Tonga); halato? (Tahiti); salato ni koro (Fiji); palilolia (Guam)
DISTRIBUTION: all the main island groups except Hawai'i

Erect monoecious herb up to 80 cm in height, with irritant hairs on the foliage and inflorescence. Stems pubescent, reddish at the base. **Leaves** alternate, simple, blade ovate, 3—17 x 1— 10 cm, acuminate at the apex, rounded to subcordate at the base; surfaces with imbedded cystoliths, upper surface appressed-pubescent, lower surface with scattered hairs; margins serrate; petiole 1—13 cm long. **Inflorescence** of narrow axillary panicles up to 25 cm long with flowers in fascicles. **Flowers** unisexual, apetalous, with 4 or 5 greenish sepals 0.3—1.5 mm long; stamens of male flowers 4 or 5; ovary of female flower superior, stigma 3-lobed. **Fruit** a green, compressed-ovoid achene 1.5—2 mm long, surrounded by a membranous wing that is dispersed with the achene.

Laportea interrupta (L.) Chew is a probably native to Southeast Asia, but was an ancient introduction as far east as Tahiti. It is uncommon to occasional in disturbed places especially in plantations and around houses at up to over 500 m elevation. It was once reported from Hawai'i, but is probably no longer found there or in Tahiti.

This erect herb can be distinguished by its irritant hairs, alternate, ovate leaves with coarsely toothed margins, long, narrow panicles of inconspicuous green flowers in clusters at intervals along the axis, and small, ovoid, seed-like fruits. Synonym: *Fleurya interrupta* (L.) Wight. A similar species on atolls in the Society and Tuamotu Islands and Micronesia, *Laportea ruderalis* (Forst.f.) Chew, is glabrous and has branching panicles.

PILEA MICROPHYLLA
Urticaceae (Nettle Family)

COMMON NAMES: rockweed, artillery plant
DISTRIBUTION: all the main island groups

Prostrate annual or short-lived herb. Stems succulent, glabrous, rooting from the nodes. **Leaves** opposite and mostly in unequally-sized pairs, blade variable but mostly obovate, usually 2—7 x 1—3 mm, rounded to broadly acute at the apex, acute to attenuate at the base; surfaces glabrous, but with embedded linear cystoliths; margins entire; petiole 1—4 mm long; stipules *ca.* 1 mm long. **Inflorescence** in short, axillary cymes or sessile clusters. **Flowers** unisexual, apetalous, green to white, 0.6—1 mm long, plants monoecious or dioecious. Male flowers with a 4-parted calyx and 4 stamens. Female flower with 3 calyx lobes and a superior ovary; stigma sessile. **Fruit** an ellipsoid achene 0.5—1 mm long, with a single seed that is ejected from the ripe fruit.

Pilea microphylla (L.) Liebm. is native from tropical South America to Florida, but is now widespread throughout the tropics. It was first recorded from the Pacific Islands in the 1920s (Marquesas, Hawai'i, Samoa), where it is occasional in disturbed places, especially on rock walls, sidewalks, and in road cracks.

This prostrate herb can be distinguished by its succulent stems, tiny opposite leaves often in unequally sized pairs, tiny, green, inconspicous flowers, and 1-seeded fruit that ejects its seed. A related weedy species in Hawai'i, *Pilea peploides* (Gaud.) Hook. & Arnott, differs in having larger, ovate to round leaves 3-veined from the base.

CLERODENDRUM CHINENSE
Verbenaceae (Verbena Family)

COMMON NAMES: clerodendrum; pīkake hohono (Hawai'i); losa Honolulu (Samoa)
DISTRIBUTION: Hawai'i, Samoa, Tahiti, Fiji

Woody perennial shrub 1.2—2 m in height, spreading by underground root suckers. Stems angular, puberulent when young. **Leaves** opposite, simple, blade broadly ovate, 6—20 x 6—18 cm, acute at the apex and truncate to cordate at the base; surfaces finely pubescent; margins wavy to irregularly toothed or weakly lobed; petiole mostly 2—10 cm long. **Inflorescence** a dense, terminal, subglobose, fragrant corymb 4—12 cm in diameter, bearing oblong to lanceolate bracteoles. **Calyx** campanulate, 1.2—1.8 cm long, red to purple, deeply divided into 5—8 lanceolate lobes. **Corolla** sympetalous, salverform, 2—4 cm across, deeply divided into 10 rounded, white to pink lobes. Stamens modified into the inner 5 corolla lobes. Ovary absent. **Fruit** not formed, hence the plant is sterile.

Clerodendrum chinense (Osbeck) Mabberley is native to southern Asia, possibly to China, and was first recorded from the Pacific Islands in 1864 (Hawai'i). It is locally abundant in places, such as fallow land in relative wet areas and along streams, at up to 670 m elevation. It often forms dense thickets by root suckers and is difficult to eradicate, despite its absence of seeds.

This shrub can be distinguished by its large ovate leaves, dense terminal cymes of flowers, purple to red calyx, showy white to pink, fragrant, 10-lobed corolla, and absence of fertile stamens or an ovary. Synonyms: *Clerodendrum fragrans* Hort. ex Vent., *C. philippinum* Schauer (the name used in the flora of Hawai'i, Wagner *et al.* 1990).

LANTANA CAMARA
Verbenaceae (Verbena Family)

COMMON NAMES: lantana; lākana (Hawai'i); lātana (Samoa); talatala (Tonga)
DISTRIBUTION: all the main island groups

Erect, branching shrub 0.5—2 m in height with bad-smelling foliage. Stems 4-angled, finely pubescent, prickly. **Leaves** opposite, simple, decussate, blade ovate, 1.5—10 x 1—6 cm, acute at the apex, rounded to cuneate at the base; upper surface scabrous and rugose, lower finely pubescent; margins toothed; petiole 0.3—2 cm long. **Inflorescence** a dense, axillary, flat-topped, head-like spike 1—3 cm across. **Calyx** cup-shaped, 1.2—2.2 mm long, shallowly 2-lobed, subtended by a bracteole 4—6 mm long. **Corolla** salverform, tube curved, 6—9 mm long, limb spreading, 4—8 mm across, yellow, orange, red, or pink in the same head. Stamens 4, epipetalous. Ovary superior. **Fruit** a shiny, dark purple or black, globose drupe 4—6 mm long.

Lantana camara L. is native to tropical America, but is now found throughout the tropics. It was first recorded from the Pacific Islands in 1858 (Hawai'i), where it is occasional to locally common in disturbed places, such as pastures, roadsides, and shrublands, and sometimes in native forest in Hawai'i, at up to 1000 m elevation. It is a serious pest because of its prickles, poisonous foliage, and habit of forming thickets. The leaves are used to treat cuts in Tonga (Whistler 1992b).

This shrub can be distinguished by its square stems, prickles, coarse, opposite leaves, flat-topped, head-like inflorescence of multicolored flowers, and shiny black, globose fruits.

STACHYTARPHETA JAMAICENSIS
Verbenaceae (Verbena Family)

COMMON NAMES: Jamaica vervain; ōwī, oī (Hawai'i)
DISTRIBUTION: Hawai'i, Samoa, Tonga, Guam, Belau

Low subshrub 0.6—1.2 m in height. Stems nearly glabrous, somewhat woody at the base. **Leaves** opposite, simple, blade elliptic to obovate, 1.5—8 x 1—4 cm, rounded to acute at the apex, cuneate at the base; surfaces mostly glabrous; margins serrate; petiole 0.5—3 cm long, narrowly winged. **Inflorescence** a terminal spike 15—50 cm long, 2—4 mm thick, with embedded solitary flowers subtended by a lanceolate bract 4.5—7 mm long. **Calyx** tubular, 4.5—7 mm long, shallowly 5-lobed. **Corolla** salverform, tube curved, 8—11 mm long, limb spreading, 6—9 mm across, pale blue to lavender. Stamens 2, epipetalous. Ovary superior. **Fruit** an oblong nutlet 4—5.5 mm long, style base persistent, nutlet enclosed within the persistent calyx, splitting into 2 black segments.

Stachytarpheta jamaicensis (L.) Vahl is native to tropical America, but is now widespread in the tropics. It was first recorded from the Pacific Islands in 1913 (Hawai'i), where it is common in dry disturbed places, such as *Leucaena* scrub forest and roadsides, especially in Hawai'i, at up to 430 m elevation.

This subshrub can be distinguished by its low habit, opposite leaves, serrate leaf margins, long, thick spike with embedded flowers covered by a lanceolate bract, and pale blue to lavender corollas with a spreading limb. A similar species in Hawai'i, *Stachytarpheta dichotoma* (Ruiz & Pavon) Vahl, often misnamed *S. australis* and *S. cayennensis*, differs in being more erect, and having thinner spikes and hairy leaves. See also *Stachytarpheta urticifolia* (p. 140).

STACHYTARPHETA URTICIFOLIA
Verbenaceae (Verbena Family)

COMMON NAMES: blue rat's-tail; mautofu (Samoa); hiku ʻi kumā (Tonga)
DISTRIBUTION: all the main island groups

Widely branching subshrub 0.5—1.5 m in height. Stems sparsely puberulent, often purple, somewhat woody at the base. **Leaves** opposite, simple, ovate to elliptic, 3—10 x 2—5 cm, acute at the apex, cuneate to attentuate at the base; surfaces glabrous, upper rugose; margins coarsely dentate; petiole 0.3—2 cm long, winged. **Inflorescence** of terminal and axillary spikes 15—50 cm long, 2—3 mm thick, with flowers embedded in the rachis with an acuminate-tipped bract 3.5—5.5 mm long. **Calyx** tubular, 4.5—6.5 mm long, 5-lobed. **Corolla** salverform, tube 7—9 mm long, limb 8—12 mm across, purple to dark blue. Stamens 2, epipetalous. Ovary superior. **Fruit** an oblong nutlet 3—5 mm long, enclosed in the calyx, splitting into 2 black segments, style base persistent.

Stachytarpheta urticifolia (Salisb.) Sims is native to tropical America and was first recorded from the Pacific Islands in 1893 (Samoa). It is common to locally abundant in disturbed places, such as roadsides, fallow land, pastures, and plantations, at up to 850 m elevation.

This subshrub can be distinguished by its opposite, coarsely dentate, rugose leaves, spikes with flowers embedded in the rachis under an attenuate-tipped bract, and purple to blue, sympetalous corollas with a spreading limb. Synonyms: *Stachytarpheta indica* and *S. jamaicensis* of many authors. Spelled *S. urticaefolia* in the flora of Fiji (Smith 1991). A similar species from Samoa, Tahiti, and the Marianas, which may be *Stachytarpheta cayennensis* (Ruiz & Pavon) Vahl, is a spreading shrub with drooping spikes and white flowers.

VERBENA LITORALIS
Verbenaceae (Verbena Family)

COMMON NAMES: verbena; ōwī, oī (Hawai'i)
DISTRIBUTION: Hawai'i

Slender subshrub up to 1.5 m in height. Stems mostly glabrous, 4-angled, somewhat woody at the base. **Leaves** opposite, simple, decussate, blade lanceolate to oblanceolate, 3—12 x 0.7—3.5 cm, acute at the apex, attenuate at the base; surfaces finely scabrous, prominently veined; margins coarsely dentate; petiole 0—5 mm long, winged. **Inflorescence** of loosely branching, terminal and axillary panicles of several narrow, many-flowered spikes 3—12 cm long. **Calyx** tubular, 1.5—2.3 mm long, shallowly 5-toothed, subtended by a lanceolate bract 1.5—2.5 mm long. **Corolla** salverform, 3—4 mm long, with 5 rounded lobes, lavender to blue. Stamens in 2 pairs, epipetalous. Ovary superior. **Fruit** ovoid, 1.2—1.8 mm long, composed of four 1-seeded nutlets that break apart.

Verbena litoralis Kunth is native to tropical America and was first recorded from the Pacific Islands in 1837 (Hawai'i). It is occasional to common in disturbed places, such as roadsides and trailsides in Hawai'i, as well as in open native forest, at up to 2200 m elevation. The plant is used in native remedies in Hawai'i (Whistler 1992b).

This subshrub can be distinguished by its square stems, leaves in opposite pairs at right angles to the pair above and below, loosely branching panicles of narrow, many-flowered spikes, and pale blue to lavender corolla. A related species uncommon in Hawai'i and occasional in Fiji, *Verbena bonariensis* L., differs in having leaves with a cordate base clasping the stem, shorter spikes, and larger corolla (over 7 mm long).

VITEX PARVIFLORA
Verbenaceae (Verbena Family)

COMMON NAMES: none
DISTRIBUTION: Guam and Belau

Medium-sized tree up to 10 m or more in height. Stems finely pubescent, marked with white lenticels. **Leaves** opposite, trifoliate, leaflets short-stalked, blade elliptic to ovate, 5—15 x 2—5 cm, attenuate at the apex, acute to rounded at the base; margins entire; surfaces glabrous; petiole 2—10 cm long. **Inflorescence** of axillary and terminal, many-flowered panicles 10—20 cm long. **Calyx** cup-shaped, 1—1.5 mm long, finely pubescent, margin entire. **Corolla** bilabiate, 4—7 mm long, violet, upper lip 2-lobed, lower lip 3-lobed. Stamens in 2 unequal pairs. Ovary superior, 4-celled. **Fruit** a black, globose drupe 5—8 mm in diameter, subtented by the saucer-shaped calyx 5—7 mm across.

Vitex parviflora Juss. is native to the Philippines, and was first recorded from the Pacific Islands in 1954 (Guam). It is occasional to common in disturbed places such as *tangantangan* scrub forest, as well as in limestone forest on Guam, and is less common in Belau.

This tree can be distinguished by its oppositely arranged leaves, long petioles, blades divided into three ovate to elliptic leaflets, many-flowered axillary and terminal panicles, violet, 2-lipped corollas, and black drupes subtended by a saucer-shaped calyx. It is related to *Vitex trifolia*, a native littoral tree or shrub that is widespread in the Pacific Islands (Whistler 1992a).

AGAVE SISALANA
Agavaceae (Century Plant Family)

COMMON NAMES: sisal; malina (Hawai'i)
DISTRIBUTION: Hawai'i, Fiji (cultivated elsewhere)

Perennial plant lacking a stem. **Leaves** forming a large rosette, simple, blade linear-lanceolate, up to 150 x 10 cm, apex with a dark brown spine 2—2.5 cm long, base subsessile; surfaces bluish and glaucous when young, green at maturity; margins entire or with numerous prickles 2—4 mm long. **Inflorescence** a tall erect, subwoody panicle up to 10 m in height, forming after the plant is 8—20 years old, after which the plant dies. **Calyx** with an urceolate tube 1.5—2 cm long and 3 greenish white sepals 4.5—6 cm long. **Corolla** with petals similar to sepals. Stamens 6, exserted. Ovary inferior. **Fruit** a capsule *ca.* 6 cm long, but rarely forming; instead, the panicle produces numerous bulbils in the bracteoles after flowering.

Agave sisalana Perrine is native to Mexico, but is now widely cultivated. It was first reported from the Pacific Islands in 1893 (Hawai'i), where it was originally introduced as a source of fiber (sisal hemp), but is now naturalized in Hawai'i and Fiji in dry, disturbed places, often on hills and in scrub vegetation at up to 450 m elevation.

This large succulent plant can be distinguished by its basal rosette of large leaves bluish and glaucous when young, black, shiny, spine-like leaf tips, large subwoody flowering stalk up to 10 m in height occurring only at maturity, and numerous bulbils produced on the inflorescence. Synonym: *Agave rigida* Mill. var. *sisalana*. A similar species in Hawai'i, *Furcraea foetida* (L.) Haw., differs in having bright green leaves with a blunt, green tip.

COMMELINA BENGHALENSIS
Commelinaceae (Spiderwort Family)

COMMON NAMES: hairy honohono (Hawai'i); kaningi (Tonga)
DISTRIBUTION: Hawai'i, Samoa, Tonga, Guam

Weak-stemmed herb branched from the base. Stems up to 40 cm long, striate, densely pubescent, rooting at the nodes. **Leaves** alternate, simple, blade ovate to suborbicular, 2—5 x 1.2—3.5 cm, acute to rounded at the apex, rounded at the base; surfaces parallel-veined, pubescent; margins entire; petiole winged and extending into the sheath, with red hairs on the margin. **Inflorescence** a cyme opposite a leaf, protruding from a folded, leaf-like, suborbicular spathe 1—1.5 x 1.6—2.8 cm, fused at the base; peduncle 5—15 mm long; underground flowers also often produced. **Calyx** of 3 unequal sepals 2—4 mm long. **Corolla** of 3 unequal petals 4—6 mm long, the upper two reniform, the lower one spathulate, smaller. Stamens 3. Ovary superior. **Fruit** an ovoid capsule 4—6 mm long.

Commelina benghalensis L. is native to tropical Asia and Africa, and was first recorded from the Pacific Islands in 1904 (Samoa). It is occasional in relatively dry, disturbed places, such as croplands and roadsides, at up to 200 m elevation. It is particularly common in Tonga, where it is an agricultural pest because it often reroots after being dug up.

This weak-stemmed herb can be distinguished by its pubescent, ovate leaves, strongly veined leaf sheaths, petioles 3—10 mm long with conspicuous red hairs on the margin, and blue, 3-petaled flowers protruding from leaf-like spathes. It differs from *Commelina diffusa*, which has narrower spathes, subsessile leaves, and lacks the reddish hairs on the petiole.

COMMELINA DIFFUSA
Commelinaceae (Spiderwort Family)

COMMON NAMES: commelina, dayflower; honohono (Hawai'i); mau'utoga (Samoa)
DISTRIBUTION: all the main island groups

Annual erect to decumbent herb. Stems up 40 cm long, glabrous, rooting at the nodes. **Leaves** alternate, simple, blade lanceolate, 3—10 x 1—5 cm, acute to acuminate at the apex, rounded to cuneate at the base, subsessile; surfaces glabrous; margins entire; sheath 1—2.8 cm long, ciliate on the upper edge. **Inflorescence** a few-flowered cyme opposite a leaf, enclosed or protruding from a cordate, leaf-like spathe 1.5—3 cm long; peduncle 0.5—2.5 cm long. **Calyx** of 3 lanceolate to ovate, unequal sepals 2—5 mm long. **Corolla** of 3 unequal blue petals 5—9 mm long, upper 2 reniform, lower one ovate, smaller. Stamens 3, staminodes 3, free. Ovary superior, 3-celled. **Fruit** an ovoid capsule 6—10 mm long, enclosed within the spathe, containing 5 pitted seeds.

Commelina diffusa Burm. f. is probably native to tropical Asia, but was an ancient introduction to the islands as far east as Samoa, and a European introduction in Hawai'i and elsewhere. It is occasional to locally abundant in relatively wet, disturbed places, such as taro patches, roadside ditches, and disturbed forest, at up to 1200 m elevation, especially in Samoa. It can be a troublesome weed in wetland crops.

This weak-stemmed herb can be distinguished by the lanceolate, subsessile leaves with parallel veins, strongly nerved leaf sheaths around the stem, and blue, 3-petaled flowers protruding from folded, heart-shaped bracts. Synonym: *Commelina nudiflora* of some authors.

CYPERUS COMPRESSUS
Cyperaceae (Sedge Family)

COMMON NAMES: none
DISTRIBUTION: all the main island groups

Tufted annual sedge with fibrous roots. Culms erect, 4—60 cm in height, 3-sided, glabrous. **Leaves** few, basal, blade linear, flat, 1—3 mm in diameter, shorter than the culms; leaf sheath membranous, pale brown, striate. **Inflorescence** umbelliform with 3—12 spikelets on 2—5 rays up to 8 cm long, or sessile in dense clusters, subtended by 2—4 unequal, leaf-like bracts 4—20 cm long. **Spikelets** 3—12 per axis, lanceolate to oblong, 0.5—2.5 cm long, 15—40-flowered, laterally compressed, imbricate, green turning yellow. Glumes ovate, 3—3.5 mm long, strongly folded with an acute keel, acute at the apex with a short mucro. Stamens 3. Ovary superior, style 3-lobed. **Fruit** a broadly obovate, 3-sided achene 1—1.3 mm long, shiny dark brown.

Cyperus compressus L. is of unknown origin, but is now pantropic in distribution. It was first recorded from the Pacific Islands in *ca.* 1904 (Samoa), where it is occasional in disturbed places such as roadsides, waste areas, and wetland margins at up to 300 m elevation.

This sedge can be distinguished by its fibrous roots and flattened, imbricate, narrowly lanceolate, yellow to green, many-flowered spikelets in dense, sessile clusters of 3—12 or in umbels on rays up to 8 cm long. Synonym: *Chlorocyperus compressus* (L.) Palla. Many species of *Cyperus* are found in the islands. A similar species common in lawns and shady places in Hawai'i, *Cyperus gracilis* R. Br., differs in having thinner stems and fewer (2—6) and shorter (5—12 mm long) spikelets.

CYPERUS ROTUNDUS
Cyperaceae (Sedge Family)

COMMON NAMES: nut sedge; kili'o'opu (Hawai'i); mumuta (Samoa); pakopako (Tonga)
DISTRIBUTION: all the main island groups

Perennial sedge 10—60 cm in height. Culms mostly solitary, 3-angled, glabrous, arising from subglobose, blackish, scaly tubers produced at the ends of long, wiry, underground stolons. **Leaves** few, basal, linear, shorter than the culm, 2—5 mm wide, folded along the midrib; leaf sheath brown, disintegrating into fibers. **Inflorescence** a loose cluster of 2—8 slender, unequal rays 0.5—8 cm long bearing 2—8 spikelets and subtended by 2 or 3 unequal bracts 1—12 cm long. **Spikelets** 3—10 per ray, linear, laterally compressed, 0.7—3 cm long, red-brown, 10—30-flowered, imbricate, loosely arranged, ascending to spreading. Glumes several, folded, brown with a green keel, 2.5—3.2 mm long. Stamens 3. Ovary superior, stigmas 3. **Fruit** a brown, oblong, 3-sided achene 1—1.5 mm long.

Cyperus rotundus L. is of uncertain origin, but is now found throughout the tropical and temperate areas of the world. It was first recorded from the Pacific Islands in *ca.* 1850 (Hawai'i), where it is occasional to locally common in disturbed places such as roadsides, lawns, and croplands at up to 800 m elevation. Its tubers make it extremely difficult to eradicate from croplands and lawns.

This sedge can be distinguished by its spreading underground stolons and tubers, 3-angled stems, terminal spikes of linear, reddish brown, spreading to ascending spikelets 0.7—3 cm long, and 3 stigmas on the ovary. Synonym: *Chlorocyperus rotundus* (L.) Palla.

FIMBRISTYLIS DICHOTOMA
Cyperaceae (Sedge Family)

COMMON NAMES: none
DISTRIBUTION: all the main island groups

Erect annual or perennial sedge 40—90 cm in height, arising from a very short rhizome. Culms thin, tufted, 3-angled, glabrous, striate. **Leaves** basal, linear, shorter than the culms, 0.5—2 mm wide; leaf sheath tightly surrounding the culm, cylindrical, light brown. **Inflorescence** a cluster of 2—12 slender, simple or compound rays of spikelets 0.5—9 cm long, subtended by 2—5 bracts, the lowest 1 or 2 leaf-like, the others bristle-like. **Spikelets** solitary or in clusters of 2 or 3, brown, many-flowered, ovate to elliptic, round in cross-section, 4—12 mm long. Glumes many, brown, ovate, 2—3 mm long. Stamens 2. Ovary superior, stigmas 2. **Fruit** a yellowish, biconvex, obovate achene 0.8—1.3 mm long, finely pitted, with a brown knob at the base.

Fimbristylis dichotoma (L.) Vahl is of uncertain origin, but is now distributed throughout the tropics. It is either native or an ancient introduction to the Pacific Islands, where it is occasional to locally common in sunny, relatively moist disturbed places, such as lawns, cane fields, and in wetlands at up to 2700 m elevation.

This tufted sedge can be distinguished by its tall, 3-angled stems, 2—5 unequal bracts with the upper 1 or 2 leaf-like and the others bristle-like, several rays bearing one or more brown, ovate to elliptic spikelets 4—12 mm long and round in cross-section, and 2 stigmas. Synonyms: *Fimbristylis annua* of some authors, *F. diphylla* (Retz.) Vahl, *F. polymorpha* Boeck.

KYLLINGA NEMORALIS
Cyperaceae (Sedge Family)

COMMON NAMES: kili'o'opu (Hawai'i); mo'u upo'o (Tahiti)
DISTRIBUTION: all the main island groups

Perennial creeping sedge spreading by means of a long-creeping rhizome. Culms tufted or spaced, erect, up to 55 cm in height, 3-angled. **Leaves** many, usually shorter than the culm, linear, 1.5—3 mm wide; leaf sheath brown to purple-brown. **Inflorescence** a globose terminal head 5—10 mm in diameter, sometimes with 2 or 3 smaller fused lateral ones, subtended by 3 or 4 unequal, spreading, leafy bracts up to 20 cm long. **Spikelets** ovate to lanceolate, white, 2.5—3 mm long, 1-flowered. Glumes usually 5, upper ones longest, boat-shaped, white variegated with brown, with a green keel. Stamens 3. Ovary superior, stigmas 2. **Fruit** an oblong to suborbicular, lens-shaped achene 1.2—1.5 mm long, brown.

Kyllinga nemoralis (Forst.) Dandy ex Hutchinson & Dalziel is native to somewhere in the Old World tropics, but was an ancient introduction to the islands as far east as Tahiti, and a European introduction to Hawai'i. It is common to abundant in relatively moist, disturbed places such as lawns, pastures, and plantations at up to 850 m elevation.

This creeping sedge can be distinguished by its 3-angled stems, white, terminal, globose heads 5—10 mm in diameter, and winged glumes. Synonyms: *Cyperus kyllingia* Endl., *Kyllinga cephalotes* (Jacq.) Druce, *K. monocephala* Rottb. A very similar species found in all the main island groups, *Kyllinga brevifolia* Rottb. (= *Cyperus brevifolius* (Rottb.) Hassk.), differs in having green heads of spikelets with unwinged glumes.

KYLLINGA POLYPHYLLA
Cyperaceae (Sedge Family)

COMMON NAMES: Navua sedge (Samoa, Fiji)
DISTRIBUTION: Samoa, Tahiti, Fiji

Creeping perennial sedge up to 75 cm in height. Culms 3-angled, glabrous, congested on the knotty, purple-scaled rhizome to form dense clumps. **Leaves** basal, 2—4, linear and shorter than the culms, 2—4 mm wide; lower leaf sheaths leafless, surrounding the culm base. **Inflorescence** a green, subglobose head 8—15 mm in diameter, formed from 1—3 confluent spikes and subtended by 5—8 drooping, unequal, leaf-like bracts up to 15 cm long and wider than the leaves. **Spikelets** green, densely packed on the head, 1—2-flowered, laterally compressed, narrowly elliptic, 2.5—3.5 mm long. Glumes several, lanceolate to ovate, tip slightly curved. Stamens 3. Ovary superior, style elongated, 2-lobed. **Fruit** a dark, oblong to obovate, biconvex achene 1.5—2 mm long.

Kyllinga polyphylla Willd. ex Kunth is native to tropical Africa and was first recorded from the Pacific Islands in *ca.* 1942 (Fiji). It is occasional to locally abundant in relatively moist, disturbed places, such as pastures and roadsides, at up 700 m elevation. Because it spreads rapidly by underground rhizomes and is not eaten by cattle, it is a serious pest in pastures.

This sedge can be distinguished by its purple-scaled, creeping underground rhizomes, green, subglobose heads 8—15 mm in diameter, and 4—8 leaf-like bracts wider than the leaves. Synonym: *Cyperus aromaticus* Mattf. & Kükenth It is somewhat similar to the widespread *Kyllinga brevifolia*, but is much more robust and has wider bracts.

PYCREUS POLYSTACHYOS
Cyperaceae (Sedge Family)

COMMON NAMES: none
DISTRIBUTION: all the main island groups

Annual or perennial herb. Culms tufted, erect, 16—100 cm in height, 3-angled, smooth. **Leaves** few, shorter than the culms, blade linear, 1.5—3 mm wide, stiff; leaf sheath reddish brown. **Inflorescence** a loose corymb with 2—6 rays up to 7 cm long, or condensed into head-like clusters 1.5—5 cm across, subtended by 4—8 leafy, unequal bracts 1—30 cm long, the lowest usually longer than the corymb. **Spikelets** digitately arranged, linear to linear-lanceolate, 7—12 mm long, acute at the apex, flattened, 9—15-flowered, yellow-brown. Glumes oblong to ovate, 1.5—2 mm long, acute at the apex, keel green. Stamen 1. Ovary superior, stigmas 2. **Fruit** an oblong to ovoid achene 0.9—1.2 mm long, brown.

Pycreus polystachyos (Rottb.) P. Beauv. is of uncertain origin, but is now found throughout the tropical and subtropical regions of the world. It was first recorded from the Pacific Islands in 1888 (Hawai'i, where it is listed as native), and is occasional to locally common in relatively moist, disturbed places, such as roadsides, canefields, and wetland margins, at up to 1300 m elevation.

This tufted sedge can be distinguished by its yellowish brown, linear-lanceolate spikelets 7—12 mm long in dense terminal clusters or at the ends of 2—6 rays up to 7 cm long, 4—8 leafy, unequal bracts up to 30 cm long, and 2 stigmas. Synonym: *Cyperus polystachyos* Rottb., a name still used by some authors in recent publications.

DIOSCOREA BULBIFERA
Dioscoreaceae (Yam Family)

COMMON NAMES: bitter yam; hoi (Hawai'i, Tonga); soi (Samoa)
DISTRIBUTION: all the main island groups

Scrambling or high-climbing dioecious vine with a large underground tuber. Stems up to 30 m long, glabrous, grooved and twisted in a clockwise direction, bearing brown, axillary, lumpy, subglobose aerial tubers up to 8 cm long. **Leaves** alternate, simple, blade cordate, mostly 8—25 x 6—22 cm, acuminate to caudate at the apex, cordate with a broad sinus at the base; surfaces glabrous, 11- or 13-nerved from the base; margins entire; petiole 4—15 cm long. **Inflorescence** of 2—6 thin, pendulous, axillary, unisexual spikes or panicles of spikes 5—15 cm long. **Calyx** of 3 lanceolate, white sepals 1.5—2.5 mm long. **Corolla** similar to calyx. Stamens 6 in male flowers. Ovary of female flower inferior. **Fruit** a brown, papery, oblong, 3-winged capsule 2—3 cm long, with winged seeds.

Dioscorea bulbifera L. is native to somewhere in the Old World tropics and was an ancient introduction to the Pacific Islands as far east as Hawai'i. It is occasional to locally abundant in relatively moist, disturbed places, such as plantations and secondary forest, at up to 900 m elevation, especially in Samoa. Its edible tuber is used only as a famine food, since it contains a bitter substance that must be removed by washing and cooking.

This herbaceous vine can be distinguished by its alternate, palmately veined, heart-shaped leaves, aerial tubers often in the leaf axils, stems that twist in a clockwise direction (looking at the tip of the stem), several axillary spikes or panicles of spikes with tiny white flowers, and brown, oblong, winged capsules.

AXONOPUS COMPRESSUS
Poaceae (Grass Family)

COMMON NAMES: carpetgrass
DISTRIBUTION: all the main island groups

Creeping perennial grass. **Culms** erect or prostrate, 15—60 cm long, laterally compressed, rooting at the hairy nodes. **Leaf sheath** compressed, keeled, hairy along margins; ligule a fringed membrane *ca.* 0.5 mm long. **Leaf blade** lanceolate, flat, 4—20 x 0.3—1 cm, blunt at the apex, broadly rounded at the base; surfaces glabrous or upper surface sparsely hairy, margins mostly ciliate. **Inflorescence** a panicle of 2—4 racemes 4—8 cm long, usually 2 at the top and one below; axis long and thin. **Spikelets** oblong, acute, *ca.* 2 mm long, green or purple, in 2 rows on one side of the rachis. Lower glume absent, upper glume as long as spikelet, with 2 nerves near each margin; sterile lemma similar to it.

Axonopus compressus (Sw.) P. Beauv. is native to tropical America and was first recorded from the Pacific Islands in 1923 (Guam). It is common in pastures, along roadsides, and in other disturbed, relatively wet places, often dominating lawns, at up to 800 m elevation. It has recently been reported from Hawai'i.

This creeping grass can be distinguished by its blunt-tipped leaves with ciliate margins, panicles typically with 3 narrow racemes (two at the top and one slightly below), and oblong spikelets in two rows on one side of the rachis. The similar *Axonopus fissifolius* (Raddi) Kuhlm. (=*A. affinis* Chase), which differs in its narrower leaves, glabrous nodes, and grain that fills the whole spikelet, is present in Hawai'i, Tahiti, and Fiji.

BOTHRIOCHLOA PERTUSA
Poaceae (Grass Family)

COMMON NAMES: pitted beardgrass
DISTRIBUTION: Hawai'i, Guam

Tuft-forming perennial grass. **Culms** erect to ascending, 30—100 cm in height, hollow, glabrous, rooting at the bearded lower nodes. **Leaf sheath** keeled, hirsute; ligule a fringed membrane 0.7—1.2 mm long. **Leaf blade** linear, 3—4 mm wide, with scattered hairs at the base and on the margins. **Inflorecence** of 3—7 subdigitate, often purplish, ascending racemes 3—5 cm long, on a thin peduncle up to 12 cm long. **Spikelets** paired, lanceolate, 3—4 mm long, often purple; pedicellate spikelet sterile or staminate, on a densely bearded pedicel *ca*. 3 mm long, the other spikelet bisexual. First glume with a conspicuous pit in the middle, often bidentate at apex; second glume as long as spikelet; sterile lemma with a slender, geniculate, reddish brown awn 15—20 mm long.

Bothriochloa pertusa (L.) A. Camus is native to the Old World tropics and was first recorded from the Pacific Islands in 1936 (Hawai'i). It is common to locally abundant in open, disturbed sites such as pastures, roadsides, and lawns at up to 1300 m elevation.

This grass can be distinguished by its 3—5 subdigitate, purple racemes, pitted glume, densely bearded stalk of the pedicellate spikelets, and long awn. Synonym: *Andropogon pertusus* (L.) Willd. A similar grass, *Bothriochloa bladhii* (Retz.) S. T. Blake [= *Dichanthium bladhii* (Retz.) Clayton], is somewhat larger, lacks the pit, and has a shorter awn. It is occasional in disturbed habitats in all the main island groups except Hawai'i.

BRACHIARIA MUTICA
Poaceae (Grass Family)

COMMON NAMES: California grass, para grass
DISTRIBUTION: all the main island groups

Coarse, spreading perennial grass. **Culms** decumbent to ascending, 60—250 cm long, often forming long stolons that root from the densely hairy nodes. **Leaf sheath** mostly covered with soft hairs; ligule very short, fringed with hairs 1—2 mm long. **Leaf blade** 10—30 x 0.5—1.5 cm; surfaces hairy, particularly the lower one; margin scabrous. **Inflorescence** an open panicle 12—25 cm long with 5—20 spaced, somewhat spreading, often zigzag racemes 3—9 cm long. **Spikelets** ovate to elliptic, 2.8—3.6 mm long, closely set on short pedicels, solitary or in clusters of 2 or 3, often tinged purple. Lower glume 1/3 as long as spikelet, 1-nerved; upper glume and sterile lemma both 5-nerved, as long as spikelet. Stigmas purple and conspicuous. Fruit infrequently forming.

Brachiaria mutica (Forssk.) Stapf is possibly native to North Africa, but is now pantropic in distribution. It was first recorded from the Pacific Islands in 1877 (Fiji), where it is common to abundant in wet areas along roadsides, in waste places, and in pastures at up to 1000 m elevation. It is often a pest in wetlands, such as in canals, where it may completely dominate and crowd out other species.

This large grass can be distinguished by its long-creeping stems, hairy foliage and nodes, large panicles of 5—20, often zigzag racemes, ovate to elliptic spikelets that are often purple, and conspicuous purple stigmas. Synonyms: *Brachiaria ambiguum* (Trin.) A. Camus, *Panicum ambigua* Trin., *P. barbinode* Trin., *P. muticum* Forssk., *P. purpurascens* Raddi.

BRACHIARIA SUBQUADRIPARA
Poaceae (Grass Family)

COMMON NAMES: none
DISTRIBUTION: all the main island groups except perhaps Tahiti

Low perennial grass. **Culms** decumbent, 20—60 cm long, rooting at the lower hairy nodes, erect or slanting in upper parts. **Leaf sheath** mostly long-hairy, particularly on upper margins; ligule very short, with a fringe of hairs *ca.* 1 mm long. **Leaf blade** 3—15 x 0.4—1 cm, relatively broad at the base; surfaces mostly glabrous; margin narrow, yellow, slightly scabrous. **Inflorescence** a loose panicle of 2—10 spreading to horizontal racemes 1—7 cm long. **Spikelets** oblong, acute, 3.3—4 mm long, glabrous, solitary or in pairs. Lower glume less than 1/2 as long as spikelet, obtuse or acute, enclosing spikelet, 5—7-nerved, green or tinged with purple; upper glume and sterile lemma as long as spikelet, 5—7-nerved, similar to each other.

Brachiaria subquadripara (Trin.) Hitchc. is native to the Old World tropics and was first recorded from the Pacific Islands in 1927 (Fiji). It is common in waste places, cultivated land, lawns, and roadsides in the lowlands.

This low, creeping grass can be distinguished by its glabrous foliage, 2—10 spreading racemes, oblong, blunt-tipped spikelets, and lower glume half as long as the spikelet and enclosing its base. Synonym: *Panicum subquadriparum* Trin. A similar species found on all the main island groups except Hawai'i, *Brachiaria paspaloides* (Presl) C. E. Hubb., differs in having acute-tipped spikelets with a longer lower glume that does not enclose the base.

CENCHRUS CILIARIS
Poaceae (Grass Family)

COMMON NAMES: Buffelgrass
DISTRIBUTION: Hawai'i, Tahiti, Fiji

Perennial grass. **Culms** ascending, 10—150 cm in height, sometimes many-branched from the lower or basal nodes to form large clumps or tussocks. **Leaf sheath** keeled, glabrous or sparsely pilose; ligule a fringe of hairs *ca.* 1 mm long. **Leaf blade** 6—25 x 0.2—0.7 cm, glabrous, or sparingly pilose at base. **Inflorescence** a cylindrical, yellow to purple panicle 6—12 cm long, with burs 4—6 mm long (excluding bristles); bristles basally fused to form a small disk, outer ones filiform, *ca.* as long as spikelet, inner ones 4—8 mm long, densely pilose on lower portion, one stouter than the rest. **Spikelets** lanceolate, 1—4 per bur, 2.3—3.3 mm long. Lower glume 1-nerved, about 1/2 as long as spikelet, upper glume and lemmas as long as spikelet, 1—3 nerved.

Cenchrus ciliaris L. is native to Africa and tropical Asia, and was first recorded from the Pacific Islands in 1932 (Hawai'i). It is common to abundant in dry, disturbed places in the lowlands at up to 800 m elevation, especially in Hawai'i, where it is the dominant grass in the dry lowland areas of all the main islands.

This clumped grass can be distinguished by its dense, cylindrical, yellow to purple panicle, soft-spiny burs 4—6 mm long, and bristles up to 8 mm long fused at the base, one stouter than the rest. Synonyms: *Pennisetum cenchroides* (L.) Rich., *P. ciliare* (L.) Link. Similar to *Pennisetum polystachion* (see p. 180), which differs in having awns up to 15 mm long with none stouter than the rest.

CENCHRUS ECHINATUS
Poaceae (Grass Family)

COMMON NAMES: sandbur; 'ume'alu (Hawai'i); vao tuitui (Samoa); hefa (Tonga)
DISTRIBUTION: all the main island groups

Annual, somewhat tufted grass. **Culms** 20—70 cm long, the lower parts often reddish, prostrate, rooting at the nodes. **Leaf sheath** keeled, glabrous or slightly hairy; ligule a dense fringe of hairs *ca.* 1 mm long. **Leaf blade** 5—30 x 0.3—1 cm; surfaces slightly hairy, smooth on lower surface, somewhat rough on upper; margins finely scabrous. **Inflorescence** a dense cylindrical spike-like raceme 3—8 cm long, bearing 10—25 spiny burs along its zigzag axis; burs purplish or straw-colored with age, globular, 3—6 mm in diameter; spines numerous, outer ones to 5 mm long, irregularly arranged, inner ones united at base to form a cup that surrounds the spikelets. **Spikelets** 2 or more per bur, 5—7 mm long.

Cenchrus echinatus L. is native to tropical America, but is now widespread in tropical regions. It was first recorded from the Pacific Islands in 1867 (Hawai'i), where it is common in disturbed and cultivated areas at low elevations (rarely up to 900 m), often on poor soil. Because of its sharp-spined bur that readily adheres to clothing or animal fur, the plant is a serious pest.

This grass can be distinguished by its spike-like racemes of hard, sharp-spined burs that cling to clothing. Other similar species, such as *Cenchrus ciliaris* (see p. 157), have bristles rather than spines.

CHLORIS BARBATA
Poaceae (Grass Family)

COMMON NAMES: swollen fingergrass; mau'u lei (Hawai'i)
DISTRIBUTION: all the main island groups

Tufted annual grass. **Culms** 25—80 cm long, erect to decumbent, rooting at the purple lower nodes, glabrous. **Leaf sheath** glabrous or with hairs at the top, keeled, often purplish, lower ones crowded; ligule a ridge *ca.* 0.5 mm high, fringed with short hairs. **Leaf blade** 5—20 x 0.1—0.5 cm, shorter upwards; surfaces glaucous, glabrous or with long, scattered hairs on upper surface near base; margins and lower midrib scabrous. **Inflorescence** a whorl of 4—15 digitately arranged, ascending racemes 2—8 cm long, purple. **Spikelets** densely overlapping, 2—2.5 mm long, obovate, arranged in 2 rows. Glumes not falling with rest of spikelet, lower one 1—1.5 mm long, upper 1.5—2.5 mm. First lemma with an awn 5—10 mm long, bearded at the tip, sterile lemmas 2, each with an awn *ca.* 5 mm long.

Chloris barbata (L.) Sw. is native to tropical America and was first recorded from the Pacific Islands in 1902 (Hawai'i). It is common in dry, disturbed places, such as roadsides, vacant lots, and pastures, mostly in the lowlands, but occasionally up to 600 m elevation. It is particularly common in the coastal areas and lowlands of Hawai'i.

This grass can be distinguished by its 4—15 spreading, purple racemes in one whorl, with the short-awned spikelets bearing persistent glumes densely arranged on the racemes. Synonyms: *Chloris inflata* Link., *C. paraguayensis* Steud. A similar species found in Hawai'i, the Society Islands, and Saipan, *Chloris virgata* Sw., differs in having white to yellowish brown racemes and only 2 (rather than 3) distinct awns.

CHLORIS RADIATA
Poaceae (Grass Family)

COMMON NAMES: radiate fingergrass
DISTRIBUTION: Hawai'i, Samoa (rare), Tonga (rare), Guam, Belau

Loosely tufted annual grass. **Culms** 25—75 cm long, usually branched and decumbent at the base, glabrous. **Leaf sheath** usually keeled, glabrous or with long hairs on the margins; ligule membranous, 0.5—1 mm long, rounded and fringed. **Leaf blade** 5—15 x 0.2—0.6 cm, shorter upwards; surfaces bluish green, upper surface with scattered long hairs; margins scabrous. **Inflorescence** of 10—20 digitately arranged, slender, erect to ascending racemes 3—7 cm long, silvery or light green. **Spikelets** densely overlapping, narrowly lanceolate, 2.6—3.2 mm long, in 2 rows. Glumes narrow, acuminate, keeled, persistent, upper one as long as the spikelet, lower one half as long. Fertile lemma as long as the spikelet, not bearded at tip, awn 7—10 mm long; sterile lemma shorter, awn 3—5 mm long.

Chloris radiata (L.) Sw. is native to tropical America and was first recorded from the Pacific Islands in 1895 (Hawai'i). It is occasional to common in disturbed places, such as roadsides and vacant lots in relatively dry areas, especially in Hawai'i, at up to 200 m elevation.

This grass can be distinguished by its 10—20 erect to ascending, silvery to green racemes 3—7 cm long, first lemma not densely bearded, and closely arranged spikelets with persistent glumes. A related species found in Hawai'i, Tonga, and Fiji, *Chloris divaricata* R. Br., differs in having fewer racemes that spread widely, looking more like a robust *Cynodon dactylon* with awns.

CHRYSOPOGON ACICULATUS
Poaceae (Grass Family)

COMMON NAMES: golden beardgrass; mānienie ‘ula (Hawai’i); inifuk (Guam)
DISTRIBUTION: all the main island groups

Creeping perennial grass forming mats by means of leafy stolons. **Culms** erect to ascending, 10—35 cm long, leaves mostly crowded near the base. **Leaf sheath** striate, mostly glabrous, old ones covering the stolons; ligule a very short ridge *ca.* 0.2 mm long. **Leaf blade** linear, 3—8 x 0.3—0.7 cm; surfaces glabrous; margins often wavy. **Inflorescence** a loose panicle 3—8 cm long, with many erect to ascending, mostly whorled branches up to 2 cm long. **Spikelets** 3, purple. Sessile spikelet 1, lanceolate, 3—4 mm long with a densely brown-hairy reflexed beard; glumes subequal, as long as spikelet, upper one with a short awn; second lemma with an awn 4—8 mm long. Pedicellate spikelets 2, lanceolate, acuminate, 5—6 mm long, staminate, on a pedicel 3—4 mm long.

Chrysopogon aciculatus (Retz.) Trin. is native to tropical Asia and probably also into the Pacific Islands, at least as far as western Polynesia, but is a European introduction to Hawai’i and Tahiti. It is locally common in lawns, streambeds, and fernlands, often on dry, poor soil, at up to 500 m elevation. Its awned and bearded spikelets readily attach themselves to fur and can cause sores on animals.

This mat-forming grass can be distinguished by its stolons covered by old leaf sheaths, erect, long-stalked panicles with many whorled branches, purple, mostly glabrous spikelets (1 sessile and awned, 2 pedicellate and acuminate), and a reflexed “beard” at the base of the sessile spikelet. Synonyms: *Andropogon aciculatus* Retz., *Rhaphis aciculata* (Retz.) Desv.

CYNODON DACTYLON
Poaceae (Grass Family)

COMMON NAMES: Bermuda grass; mānienie (Hawai'i)
DISTRIBUTION: all the main island groups

Creeping perennial grass spreading by means of underground stolons. **Culms** erect or ascending, 5—40 cm long, wiry, smooth, sometimes reddish, readily rooting at the glabrous nodes. **Leaf sheath** mostly glabrous, persistent and scarious on old stolons; ligule a fringe of hairs 0.2—0.5 mm long. **Leaf blade** 2—12 x 0.1—0.3 cm; surfaces glabrous or slightly hairy. **Inflorescence** a digitate panicle of 3—7 racemes in a single whorl, 3—7 cm long, sometimes purplish. **Spikelets** elliptic, 2—2.8 mm long, in 2 rows overlapping and appressed to the rachis. Glumes lanceolate, acuminate or with shortly awned tip, mostly shorter than spikelet, persistent after rest of spikelet falls. Lemmas ovate, as long as spikelet, keeled.

Cynodon dactylon (L.) Pers. is possibly native to tropical Africa, but is now cultivated and naturalized throughout the tropics and subtropics. It was first recorded from the Pacific Islands in Hawai'i, Samoa, and Tonga before 1840, and is common in lawns, roadsides, and other sunny, disturbed places, often being dominant in coastal areas, mostly in the lowlands, but occasionally up to 2200 m elevation.

This mat-forming grass can be distinguished by its creeping stolons, digitate inflorescences of 3—7 narrow branches 3—7 cm long, and awnless spikelets with persistent, lanceolate glumes that remain on the rachis when the rest of the spikelet falls.

DACTYLOCTENIUM AEGYPTIUM
Poaceae (Grass Family)

COMMON NAMES: Beach wiregrass
DISTRIBUTION: all the main island groups

Annual grass often spreading to form mats. **Culms** prostrate to ascending, 20—100 cm long, rooting at the glabrous lower nodes. **Leaf sheath** mostly glabrous, flattened; ligule 0.5—1 mm long, membranous with a jagged edge ending in hairs. **Leaf blade** 3—8 x 0.3—0.6 cm; surfaces sparsely covered with hairs having swollen bases, particularly near the ligule; margins scabrous. **Inflorescence** of (1—) 2—5 terminal, digitately arranged, 1-sided spikes 1—4 x 0.4—0.8 cm. **Spikelets** broadly ovate, 3—4 mm long, crowded in several rows, 3—5-flowered. Glumes *ca.* half as long as spikelet, lower one acute and persistent on the rachis, upper one short-awned and deciduous. Lemmas membranous and usually short-awned.

Dactyloctenium aegyptium (L.) Willd. is native to somewhere in the Old World tropics and was first recorded from the Pacific Islands in 1909 (Hawai'i). It is occasional in disturbed places, particularly in dry coastal areas and on beaches, sometimes at up to 250 m elevation.

This mat-forming grass can be distinguished by its digitate inflorescences of 2—5 (rarely 1), short, one-sided spikes, and flattened spikelets with the upper one bearing a short awn. It is superficially similar to *Eleusine indica* (see p. 168), which has much longer spikes, one of them usually attached below the others.

DIGITARIA CILIARIS
Poaceae (Grass Family)

COMMON NAMES: Henry's crabgrass; kūkaepua'a (Hawai'i)
DISTRIBUTION: all the main island groups

Tufted or creeping annual grass. **Culms** ascending, 20—100 cm long, base decumbent, rooting at the lower hairy or glabrous nodes. **Leaf sheath** keeled, bearing long tubercle-based hairs or sometimes glabrous, margins membranous; ligule membranous, truncate, 1—3 mm long. **Leaf blade** 4—25 x 0.3—1 cm; surfaces glabrous or hairy; one margin often crinkled. **Inflorescence** of 4—12 spreading, subdigitate racemes mostly 7—15 cm long, in 1—3 whorls; rachis 3—angled with scabrous margins. **Spikelets** lanceolate, 2.5—3.3 mm long, in unequally pedicellate pairs. Lower glume reduced to a scale, upper glume 1/2 —3/4 as long as spikelet, acute, 3-nerved, margins hairy. Sterile lemma as long as spikelet, 7-nerved, with lateral nerves crowded towards the margins, leaving broad interspaces along midvein.

Digitaria ciliaris (Retz.) Koeler is native to tropical Asia, and was first recorded from the Pacific Islands in 1909 (Hawai'i). It is common to locally abundant in sunny waste places, cultivated areas, pastures, and along roadsides at up to 900 m elevation.

This grass can be distinguished by its 4—12 spreading racemes mostly 7—15 cm long, paired, unequally stalked lanceolate spikelets, and a scale-like lower glume. Another widespread species, *Digitaria radicosa* (Presl) Miq., differs in being a more delicate species with 2—4 racemes 3—10 cm long. A third species common to abundant in Samoa, *Digitaria horizontalis* Willd., has scattered, tubercle-based hairs 3—6 mm long on its racemes.

DIGITARIA INSULARIS
Poaceae (Grass Family)

COMMON NAMES: sourgrass
DISTRIBUTION: Hawai'i, Guam

Tufted perennial grass. **Culms** erect, 100—150 cm tall with a hard, scaly, pubescent base. **Leaf sheath** much longer than internodes, usually with tubercle-based hairs 5—10 mm long, striate, compressed, keeled; ligule membranous, 2.5—5 mm long. **Leaf blade** lanceolate, 15—50 x 1—1.5 cm; upper surface scabrous, lower surface glabrous, midrib conspicuous. **Inflorescence** a narrow panicle 15—25 cm long with numerous slender, somewhat nodding, silky-pubescent branches 7—15 cm long. **Spikelets** paired, narrowly elliptic, 3.5—4 mm long, densely yellowish brown silky-pubescent. Upper glume ovate, scale-like, 0.5—1 mm long; lower glume as long as spikelet, 3—5-nerved, densely silky-pubescent with the hairs longer than it. Sterile lemma similar to upper glume.

Digitaria insularis (L.) Mez ex Ekman is native to tropical America, but is now widely naturalized in Malaysia and elsewhere. It was first recorded from the Pacific Islands in 1924 (Hawai'i), where it is common to abundant, especially in Hawai'i, on roadsides, abandoned fields, and waste places at up to 300 m elevation.

This large grass can be distinguished by its dense panicles of long, somewhat nodding racemes, and the silky spikelets in unequally stalked pairs. It is quite different from any other *Digitaria* species in the region, and until recently was put into another genus, *Tricachne*. Synonym: *Tricachne insularis* (L.) Nees.

DIGITARIA SETIGERA
Poaceae (Grass Family)

COMMON NAMES: itchy crabgrass; kukaepua'a (Hawai'i)
DISTRIBUTION: all the main island groups

Annual grass. **Culms** erect to ascending, 30—120 cm long, decumbent at the base and rooting at the lower glabrous to long-hairy nodes. **Leaf sheath** keeled, glabrous or somewhat hairy, margins membranous; ligule membranous, truncate, 1—3 mm long. **Leaf blade** 6—25 x 0.4—1.2 cm; surfaces glabrous to pubescent, with some long hairs near the ligule; one margin usually crinkled. **Inflorescence** of 3—20 subdigitate racemes scarcely spreading at maturity, mostly 5—15 cm long; rachis 3-angled, scabrous. **Spikelets** lanceolate-elliptic, 2.5—3 mm long, in unequally pedicellate pairs. Lower glume virtually absent, upper glume less than 1/3 as long as spikelet. Sterile lemma as long as spikelet, 5—7-nerved, lateral veins crowded near the edges, leaving broad, glabrous interspaces along midvein.

Digitaria setigera Roth is native, or perhaps was an ancient introduction, to the Pacific Islands and ranges from India to Hawai'i. It is common in coastal lowlands, on littoral rocks, in waste places, pastures, and along roadsides, from near sea level to 1000 m elevation. It was once used in native remedies in Hawai'i.

This grass can be distinguished by its 3—20 racemes that spread very little, spikelets 2—4 mm long, and the lower glume absent or minute. Synonyms: *Digitaria chinensis* of some authors, *D. consanguinea* Gaud., *D. microbachne* (Presl) Henr., *D. pruriens* (Trin.) Buese, *Panicum pruriens* Trin., *P. sanguinale* of some authors, *Syntherisma pruriens* (Trin.) Arthur. A similar widespread species, *Digitaria violascens* Link, differs in having 4—8 spreading racemes, spikelets less than 2 mm long, and a black grain.

ECHINOCHLOA COLONA
Poaceae (Grass Family)

COMMON NAMES: jungle rice
DISTRIBUTION: all the main island groups

Tufted annual grass. **Culms** erect or ascending, 20—80 cm long, often decumbent and reddish at the base, and rooting at the lower, mostly glabrous nodes. **Leaf sheath** keeled, compressed, glabrous; ligule absent. **Leaf blade** 4—15 x 0.3—1 cm; surfaces glabrous or slightly hairy; margins smooth or sometimes scabrous. **Inflorescence** a panicle 4—12 cm long bearing many alternating compact racemes 1—3 cm long; rachis angular and finely scabrous. **Spikelets** ovate, 2—3 mm long, short-hairy, crowded together on short pedicels. Glumes ovate, lower one *ca.* 1/2 as long as spikelet, 3-nerved, upper glume as long as spikelet, 5—7-nerved, with a short mucro. Sterile lemma similar to upper glume, fertile lemma broadly ovate and shining.

Echinochloa colona (L.) Link (sometimes misspelled *E. colonum*) is native to the Old World tropics and was first recorded from the Pacific Islands in 1835 (Hawai'i). It is common in disturbed areas, particularly in wet soil, in places such as canals and rice fields at up to 300 m elevation.

This grass can be distinguished by its panicles of short, alternating branches 3—5 mm wide, bearing densely packed, awnless, acute-tipped spikelets in several rows. Synonym: *Panicum colonum* L. A related, larger grass found in Hawai'i, Tahiti, Fiji, and Belau, *Echinochloa crus-galli* (L.) P. Beauv., differs in having acuminate or awned spikelets. Another larger grass from Samoa and Fiji, *Echinochloa stagnina* (Retz.) P. Beauv., differs in having a ligule of stiff hairs.

ELEUSINE INDICA
Poaceae (Grass Family)

COMMON NAMES: goosegrass; mānienie ali'i (Hawai'i); ta'ata'a (Samoa); takataka 'a leala (Tonga); umog (Guam)
DISTRIBUTION: all the main island groups

Tufted annual grass. **Culms** prostrate or ascending, 15—60 cm long, flattened, glabrous. **Leaf sheath** striate, flattened, keeled, slightly hairy along margins and at base; ligule membranous, 0.4—1 mm long, with a jagged edge. **Leaf blade** 6—30 x 0.3—0.8 cm, keeled; surfaces mostly glabrous; midrib and upper margins scabrous. **Inflorescence** of 2—7 terminal, one-sided spikes 4—12 x 0.3—0.7 cm, usually with one arising below the others. **Spikelets** mostly 3.5—5 mm long, glabrous, 3—7-flowered. Glumes lanceolate-ovate, 2—4 mm long, persistent after the rest of spikelet has fallen, margins membranous, keel finely scabrous. Lemmas similar to glumes.

Eleusine indica (L.) Gaertner is native to the Old World tropics and is now widely naturalized in the New World as well. It was an ancient introduction to the Pacific Islands as far east as Polynesia, but was a European introduction to Hawai'i, where it was first recorded in 1864. It is common in sunny, disturbed places, especially in lawns and along dirt roads and tracks, at up to 900 m elevation.

This grass can be distinguished by its flattened culms, 2—7 coarse, terminal, digitate branches, often with one branch attached below them, and glumes that persist on the rachis when the rest of the spikelet falls.

ERAGROSTIS TENELLA
Poaceae (Grass Family)

COMMON NAMES: lovegrass
DISTRIBUTION: all the main island groups

Delicate annual grass with a tufted base. **Culms** slender, ascending to decumbent, geniculate, 8—40 cm long; nodes dark, glabrous. **Leaf sheath** striate, mostly glabrous but pilose at the apex; ligule a fringe of hairs 0.2—0.4 mm long. **Leaf blade** linear, 4—10 x 0.2—0.6 cm; surfaces glabrous; margins smooth, sometimes rolled. **Inflorescence** an open, delicate, plumose panicle 4—15 cm long, pilose at the base of the branches. **Spikelets** oblong, 1—2 mm long, tinged with purple, 2—6-flowered. Glumes glabrous, 1-nerved, scabrid on the keel, upper glume 1/2 as long as spikelet, lower one shorter. Lemmas obtuse, imbricate; paleae conspicuously ciliate. Grain brown, falling free from the spikelet.

Eragrostis tenella (L.) P. Beauv. ex Roem. & Schult. is native to the Old World tropics and was first recorded from the Pacific Islands in 1895 (Hawai'i). It is occasional on roadsides and in sunny, disturbed places, particularly in dry localities at up to 450 m elevation.

This delicate grass can be distinguished by its feather-like, open panicle of short delicate branches, and several-flowered spikelets 1—2 mm long with persistent glumes. Synonyms: *Eragrostis amabilis* (L.) Wight & Arn. ex Hook. & Arnott, *E. plumosa* (Retz.) Link. A similar species found in Hawai'i, *Eragrostis cilianensis* (All.) Link, differs in having spikelets 5—15 mm long with 10—40 flowers. Another similar one found in Hawai'i, *Eragrostis pectinacea* (Michx.) Nees, has spikelets 3—8 mm long.

ERIOCHLOA PROCERA
Poaceae (Grass Family)

COMMON NAMES: cupgrass
DISTRIBUTION: Samoa, Tonga, Fiji, Guam

Strongly tufted annual or perennial grass. **Culms** erect to ascending, 20—80 cm long, often creeping at the base; nodes short-hairy. **Leaf sheath** striate, glabrous, keeled and compressed; ligule a short fringe of hairs *ca.* 0.5 mm long. **Leaf blade** linear, 5—20 x 0.3—0.6 cm, base broad; surfaces glabrous; margins smooth. **Inflorescence** a panicle of 4—12 flattened racemes, the lower ones 2—6 cm long. **Spikelets** lanceolate, 3—3.5 mm long, pedicellate, green or tinged with purple, with a conspicuous, swollen, purple, knob-like base. Lower glume absent, upper glume and sterile lemma similar, as long as spikelet, appressed-hairy on back, obscurely 5-nerved, tip acute or mucronate.

Eriochloa procera (Retz.) C. E. Hubb. is native to the Old World tropics and was first recorded from the Pacific Islands in 1944 (Fiji). It is occasional to locally common in the lowlands on roadsides, gardens, lawns, and other disturbed places, especially in Samoa. It is a fairly recent arrival in the islands and can be expected to spread rapidly once it becomes established, as it has in American Samoa.

This grass can be distinguished by its panicles of 4—12 flattened racemes, lanceolate spikelets 3—3.5 mm long, and particularly by the purple, knob-like base of the spikelet. Synonym: *Eriochloa ramosa* Kuntze.

EUSTACHYS PETRAEA
Poaceae (Grass Family)

COMMON NAMES: none
DISTRIBUTION: Guam, Belau

Annual tufted grass. **Culms** 20—80 cm long, erect to decumbent, compressed, often purple, rooting at the lower nodes and spreading by means of stolons. **Leaf sheath** glabrous, compressed, strongly keeled; ligule a dense fringe of hairs 0.2—0.6 mm long. **Leaf blade** linear, 4—28 x 0.3—0.8 cm, glabrous, glaucous, rounded at the tip; margins smooth. **Inflorescence** of 3—8 ascending racemes 4—11 cm long. **Spikelets** densely arranged in 2 rows, 1.8—2.2 mm long. Glumes membranous, persistent when the rest of the spikelet falls, lower one lanceolate, 0.8—1.2 mm long, upper one slightly larger, oblanceolate, bilobed at the apex, with an awn 1/2—3/4 as long as it. Lemmas ovate, dark brown, as long as spikelet, ciliate on the nerves.

Eustachys petraea (Sw.) Desv. is native to tropical America and is now also found in the southeastern U.S. It was first recorded from the Pacific Islands in *ca.* 1965 (Guam) and is common in disturbed lowland places, such as roadsides and waste areas on most of the main Micronesian islands, but has not been reported from Polynesia.

This grass can be distinguished by its compressed, often purplish culms, glaucous, round-tipped leaves, 3—8 digitately arranged spikes, glumes that persist on the inflorescence when the rest of the spikelet falls, short awn on the upper glume, and conspicuous dark brown, awnless lemmas. Synonym: *Chloris petraea* Sw. It is very similar to species of *Chloris,* but its dark brown, awnless lemmas are distinctive.

HETEROPOGON CONTORTUS
Poaceae (Grass Family)

COMMON NAMES: twisted beardgrass; pili grass (Hawai'i)
DISTRIBUTION: all the main island except Belau and Tahiti (but rare in Samoa)

Tufted perennial grass. **Culms** erect, 30—100 cm long, glabrous. **Leaf sheath** compressed, keeled, glabrous or sparsely hispid in the throat; ligule a fringed membrane 0.5—1 mm long. **Leaf blade** 10—25 x 0.3—0.7 cm; surfaces glabrous; margins finely scabrous. **Inflorescence** a solitary raceme 3—7 cm long. **Spikelets** narrowly lanceolate, 6—10 mm long, paired, one sessile and one stalked; the lower 2—6 pairs male, the upper 8—13 pairs with a sessile female or bisexual spikelet with a bearded sharp barb below it, and a stalked male or sterile one. Glumes as long as spikelet, often with tubercle-based hairs, lower glume keeled, upper one rounded on back. Fertile lemma of upper sessile flowers with a bent, flexuous awn 5—12 cm long intertwined with others.

Heteropogon contortus (L.) P. Beauv. ex Roem. & Schult. is pantropic in distribution and is native or possibly an ancient introduction to the Pacific Islands. It is occasional to locally common on lowland slopes, cliffs, and ledges at up to 700 m elevation in Hawai'i, where it is one of the few native lowland plants found in disturbed areas, but is uncommon elsewhere in the islands and perhaps extinct in some (e.g., Samoa). The barb may cause skin sores in animals. The stems were formerly used for roof thatch in Hawai'i.

This grass can be distinguished by its long raceme of paired spikelets, and especially by its long (5—12 cm) awns with their tips intertwined. Synonym: *Andropogon contortus* L.

MELINUS MINUTIFLORA
Poaceae (Grass Family)

COMMON NAMES: molasses grass; puakatau (Tonga)
DISTRIBUTION: all the main island groups (but rare in Samoa)

Tall perennial grass with sticky foliage. **Culms** erect or ascending, 50—100 cm long, profusely rooting at the long-hairy lower nodes. **Leaf sheath** densely hairy, the base of the hairs producing a sticky oil smelling like molasses; ligule a fringe of hairs 1—2 mm long. **Leaf blade** 5—20 x 0.4—1.2 cm; surfaces densely hairy, sticky. **Inflorescence** a dense, much-branched panicle 10—25 cm long, spreading or contracted, purple. **Spikelets** narrowly oblong, 1.5—2 mm long, glabrous but finely scabrous, often tinged purple. Lower glume scale-like, very short, upper glume as long as spikelet, distinctly 7-nerved, tip bilobed with a short mucro between the points. Sterile lemma similar to upper glume, distinctly 5-nerved, tip bilobed with an awn 4—10 mm long.

Melinus minutiflora P. Beauv. is native to Africa, but is now widely grown in tropical countries as a forage plant. It was first recorded from the Pacific Islands in 1905 (Samoa). It is common to locally abundant in relatively dry, open, disturbed places up to 1200 m elevation in Hawai'i, where it tends to dominate in some places, choking out seedlings and covering native herbaceous vegetation. It is less common on other islands.

This grass can be distinguished by its somewhat sticky, strong-smelling foliage, large, dense, purplish panicles, and small, oblong spikelets with an awn 4—10 mm long.

OPLISMENUS COMPOSITUS
Poaceae (Grass Family)

COMMON NAMES: basket grass
DISTRIBUTION: all the main island groups

Creeping perennial grass. **Culms** slanting to erect, 25—100 cm long, striate, mostly glabrous, creeping and forming mostly single roots at the lower hairy nodes. **Leaf sheath** striate, hairy or glabrous, ciliate on margins; ligule membranous, *ca.* 1 mm long with a fringe of hairs of similar length. **Leaf blade** lanceolate, 3—12 x 0.8—2.5 cm, with an asymmetrical base; surfaces glabrous or upper surface hairy; margins often wavy. **Inflorescence** a panicle 4—20 cm long bearing 4—12 spaced racemes, the lowest ones 2—7 cm long, upper ones progressively shorter. **Spikelets** lanceolate, 2.5—4 mm long, glabrous or hairy. Glumes *ca.* 2/3 as long as spikelet, 5—7-nerved, lower glume with a purplish awn 4—12 mm long, upper with an awn 1—3 mm long. Sterile lemma as long as spikelet, 7—11-nerved, fertile lemma smooth.

Oplismenus compositus (L.) P. Beauv. is pantropic in distribution and was probably an ancient introduction to the Pacific Islands as far east as Tahiti, and a European introduction in Hawai'i. It is common in disturbed shady places, such as open secondary forest and plantations, at up to 1750 m elevation.

This grass can be distinguished by its broad leaves, panicles of 4—12 spaced racemes 2—7 cm long, and spikelets bearing long bristles that are usually purple. It is sometimes difficult to distinguish from *Oplismenus hirtellus* (L.) P. Beauv., which is generally smaller and has shorter racemes 1—4 cm long with the spikelets more densely packed. It has a similar range, and is the more common of the two species in Hawai'i.

PANICUM MAXIMUM
Poaceae (Grass Family)

COMMON NAMES: Guinea grass; saafa (Tonga)
DISTRIBUTION: all the main island groups

Robust tufted perennial grass spreading by short rhizomes to form large bunches. **Culms** erect, stout, 1—2.5 m in height, nodes hairy. **Leaf sheath** striate, often with tubercle-based hairs, upper margins sometimes hairy; ligule membranous, 1—3 mm long with a ciliate margin. **Leaf blade** 25—75 x 0.5—2.5 cm; surfaces nearly glabrous, yellow-green; midrib prominent. **Inflorescence** a panicle 20—40 cm long, with numerous erect to spreading branches mostly 10—20 cm long, the lower ones in whorls. **Spikelets** elliptic, 2.8—3.3 mm long, green or purple, glabrous, on pedicels of varying length. Lower glume *ca.* 1/4 as long as spikelet, enclosing the base; upper glume and sterile lemma similar, as long as spikelet. Second floret rugose.

Panicum maximum Jacq. is native to Africa, but is now widely grown as a forage grass. It was first recorded from the Pacific Islands in 1871 (Hawai'i), where it is occasional to locally common in disturbed places, often dominating lowland areas in Hawai'i and elsewhere, at up to 900 m elevation.

This large grass can be distinguished by its relatively large size, yellow-green foliage, large, spreading panicles, and glabrous, awnless, elliptic spikelets with the lower glume about one quarter of its length, and with the second floret rugose. A similar, invasive wetland species found in Hawai'i and Belau, *Panicum repens* L. (torpedo grass), differs in being a shorter, rhizomatous grass with the second floret smooth.

PASPALUM CONJUGATUM
Poaceae (Grass Family)

COMMON NAMES: Hilo grass, sour paspalum, T-grass; vaolima (Samoa)
DISTRIBUTION: all the main island groups

A creeping perennial grass. **Culms** erect, 20—60 cm high, nodes glabrous, spreading by long, often reddish-purple stolons. **Leaf sheath** flattened, keeled, green or tinged with purple, margins somewhat hairy; ligule less than 1 mm long, membranous, appearing cutoff at the tip. **Leaf blade** 8—20 x 0.5—1.5 cm; surfaces glabrous to slightly hairy; margins scabrous or stiffly hairy. **Inflorescence** of two terminal, widely spreading racemes 5—15 cm long. **Spikelets** 1.5—2 mm long, ovate, flattened, pale green, the margin fringed with long silky hairs. Lower glume absent, upper glume and sterile lemmas as long as spikelet.

Paspalum conjugatum Bergius is native to tropical America and was first recorded from the Pacific Islands in 1840 (Hawai'i). It is common to abundant in lawns, roadsides, and other disturbed habitats, particularly in wet places, at up to 950 m elevation. It can be a troublesome weed in wet places such as taro patches, and in relatively moist disturbed areas it is often the dominant species.

This creeping grass can be distinguished by its flattened stems with hairy leaf sheaths, T-shaped inflorescence with 2 spreading terminal racemes, and flattened ovate spikelets with a fringed margin.

PASPALUM FIMBRIATUM
Poaceae (Grass Family)

COMMON NAMES: fimbriate paspalum
DISTRIBUTION: Hawai'i, American Samoa (rare), Tonga, Guam

Annual grass. **Culms** weakly tufted, yellowish green, erect, 30—60 cm in height, glabrous. **Leaf sheath** compressed, long-hispid on the margins, the basal ones often purple; ligule membranous, 0.6—1.5 mm long, sometimes split. **Leaf blade** flat, 8—30 x 0.5—1.5 cm, surfaces glabrous except for long hairs at the base; margins conspicuously ciliate. **Inflorescence** a panicle of 3—5 widely spaced, ascending or somewhat reflexed, alternate racemes 2—6 cm long, each with a tuft of long white hairs at the base. **Spikelets** ovate to orbicular, 1.2—1.8 mm long (excluding the fringe), tinged purple. Lower glume absent, upper glume as long as spikelet, distinctly 1-nerved, with a broad, ragged fringe up to 1 mm wide that is ciliate on its edge. Lemma as long as spikelet.

Paspalum fimbriatum Kunth is native to Central and South America, and was first recorded from the Pacific Islands in 1924 (Hawai'i). It is common in disturbed, open, relatively moist sunny places in the lowlands of Hawai'i (Kaua'i and O'ahu) at up to 280 m elevation, but is less common elsewhere in the Pacific Islands.

This grass can be distinguished by it ciliate leaf margins, panicles of 3—5 widely spaced racemes 2—6 cm long, and the purplish, flattened, conspicuously fringed spikelets. A similar species occasional in all the main island groups except Belau, *Paspalum dilatatum* Poiret, differs in having a similar number of longer (6—9 cm), narrow racemes with unfringed, green, pubescent spikelets.

PASPALUM PANICULATUM
Poaceae (Grass Family)

COMMON NAMES: none
DISTRIBUTION: all the main island groups except Hawai'i and Belau

Perennial grass. **Culms** erect to ascending, 50—100 cm long, sometimes rooting from the lower node, nodes often long-hairy. **Leaf sheath** long-hairy, striate, often longer than internodes; ligule less than 1 mm long, membranous. **Leaf blade** 10—40 x 0.8—2 cm; surfaces hairy, with a tuft of long white hairs at the base, midrib prominent beneath; margins scabrous. **Inflorescence** of 15—20 spreading to drooping racemes from an axis 10—20 cm long, the lower racemes up to 8 cm long. **Spikelets** nearly round, *ca.* 1.5 mm long, brown, crowded together in pairs on the slender, winged rachis, which often bears scattered stiff hairs. Lower glume absent, upper glume and sterile lemma similar, slightly pubescent.

Paspalum paniculatum L. is native to tropical America and was first recorded from the Pacific Islands in *ca.* 1920 (Fiji), where it was originally introduced as a pasture grass. However, it has now become one of the most common weeds, often dominating in sunny, relatively moist, disturbed places, particularly in abandoned taro plantations in Samoa, at up to 900 m elevation.

This large grass can be distinguished by its relatively large, hairy leaves, 15—20 spreading to drooping racemes, and round spikelets *ca.* 1.5 mm in diameter. A similar species occurring in Hawai'i, Samoa, Fiji, and Guam, *Paspalum urvillei* Steud., differs in having erect to ascending racemes and hairy spikelets 2—3 mm long.

PASPALUM SETACEUM
Poaceae (Grass Family)

COMMON NAMES: none
DISTRIBUTION: American Samoa, Guam, Carolines, Marshalls

Creeping perennial grass spreading by rhizomes. **Culms** slender, erect to ascending, up to 80 cm long; nodes dark, glabrous. **Leaf sheath** striate, compressed, glabrous to pubescent; ligule a narrow membrane *ca.* 0.5 mm long. **Leaf blade** linear, 4—20 x 0.4—1.2 cm, margins translucent with straight, scattered, tubercle-based hairs up to 2 mm long. **Inflorescence** an erect to arching, terminal raceme 3—8 cm long, on a thin pedicel up to 20 mm long, sometimes with a second raceme below it; 1—4 inflorescences per sheath. **Spikelets** elliptic-obovate, 1.6—2.2 mm long, glabrous or minutely pubescent, in two dense rows along the entire rachis. Lower glume absent, upper glume and sterile lemma similar to each other.

Paspalum setaceum Michx. is native to Mexico and the southeastern U.S., and was first recorded from the Pacific Islands in 1956 (Marshalls). It is common to locally abundant in lawns and disturbed places, such as roadsides, mostly in coastal and lowland areas of Micronesia and American Samoa, but is likely to spread in the region.

This creeping grass can be distinguished by its dark, glabrous nodes, long, scattered hairs on the leaf margins, 1—4 inflorescences arising from the leaf sheath, slender peduncles bearing 1 or 2 racemes 3—8 cm long, and small spikelets densely arranged in two rows. Similar to the widespread native *Paspalum orbiculare* Forst. f. (called *P. scrobiculatum* L. by some authors), which has several racemes per inflorescence and is common in wetlands.

PENNISETUM POLYSTACHION
Poaceae (Grass Family)

COMMON NAMES: feathery pennisetum
DISTRIBUTION: all the main island groups except Samoa and Tonga

Perennial grass. **Culms** loosely tufted, erect, 60—120 cm in height. **Leaf sheath** glabrous, or the lower ones sometimes pubescent; ligule a fringe of hairs *ca.* 1 mm long. **Leaf blade** 6—40 x 0.3—1.5 cm, pilose on both surfaces, base of upper surface with a tuft of silky hairs up to 6 mm long. **Inflorescence** a yellowish brown to purple, dense, cylindrical panicle 7—18 cm long. **Spikelet** lanceolate, 1.8—2.3 mm long, 1 per involucre; outer involucral bristles delicate, scabrous, 3—4 mm long, densely silky-pubescent on lower portion, one longer, 9—15 mm long. Lower glume minute or vestigial, upper glume membranous, as long as spikelet; upper lemma slightly shorter than spikelet, membranous.

Pennisetum polystachion (L.) Schult. is native to Central America and is now widely naturalized in the tropics. It was first recorded from the Pacific Islands in 1923 (Hawai'i), where it is common in disturbed, mostly dry lowland areas and cultivated fields at up to 2100 m elevation, and is especially abundant in Guam and Fiji, where it often completely dominates dry hillsides.

This grass can be distinguished by its yellowish, plumose, cylindrical, spike-like panicle 7—18 cm long, angular rachis, sessile involucres, and spikelets subtended by a ring of bristles, one of which is up to 15 mm long. Synonyms: *Cenchrus setosus* Sw., *Panicum polystachion* L., *Pennisetum setosum* (Sw.) Rich. A similar, more robust grass widespread in the Pacific is *Pennisetum purpureum* Schumach., which has a cylindrical rachis and yellow burs.

PENNISETUM SETACEUM
Poaceae (Grass Family)

COMMON NAMES: fountain grass
DISTRIBUTION: Hawai'i, Fiji

Perennial grass. **Culms** densely tufted, 20—120 cm in height, scabrous or glabrous. **Leaf sheath** 6—15 cm long, long-pilose at throat and along upper margins; ligule a dense row of silky hairs *ca.* 0.5 mm long. **Leaf blade** 40—60 x 0.1—0.2 cm, margins scabrous, in-rolled. **Inflorescence** a cylindrical, feathery, pink to purple panicle mostly 18—30 cm long; rachis unwinged. **Spikelets** lanceolate, 4.5—6.5 mm long, 1—3 in a fascicle borne on a pubescent stipe 1—3 mm long; inner involucral bristles silky-pubescent at base or scabrous, others scabrous, inner involucre up to 3 cm or more long. First glume ovate to suborbicular, up to 1/3 as long as spikelet, or absent; second glume 1/4—2/3 as long as spikelet; first lemma as long as spikelet, 3-nerved.

Pennisetum setaceum (Forssk.) Chiov. is native to North Africa, but has now been introduced and cultivated elsewhere. It was first recorded from the Pacific Islands in 1914 (Hawai'i), where it is common to abundant in dry, open places, particularly on lowland lava flows on the island of Hawai'i at up to 2100 m elevation. It has become a pest there since it out-competes native species, partly because of its adaptation to fire.

This large, clump-forming grass can be distinguished by its long, narrow leaves with in-rolled margins, reddish, feathery, cylindrical panicles up to 30 cm long, 1—3 spikelets on a fascicle borne on a short, pubescent stalk, and reddish involucre bristles up to 3 cm or more in length. Synonym: *Pennisetum ruppelii* Steud.

RHYNCHELYTRUM REPENS
Poaceae (Grass Family)

COMMON NAMES: Natal redtop
DISTRIBUTION: Hawai'i, Tonga, Tahiti, Fiji, Guam

Annual or perennial grass. **Culms** loosely tufted, erect to ascending, 30—100 cm in height, glabrous, nodes pubescent. **Leaf sheath** rounded, glabrous or sparingly to densely long-pilose, especially at the top; ligule a row of hairs *ca.* 1 mm long. **Leaf blade** 5—30 x 0.2—0.7 cm, flat or folded, glaucous, upper surface finely scabrous. **Inflorescence** a loose, ovoid to oblong panicle 6—20 cm long with ascending branches mostly 2—6 cm long. **Spikelets** ovate, 2—3.5 mm long, pink to purple but fading to silver, densely villous with long hairs extending up to 4 mm beyond apex. First glume nearly as long as spikelet, truncate or emarginate, separated from the upper glume by a short internode; upper glume as long as spikelet, usually with an awn 1—4 mm long.

Rhynchelytrum repens (Willd.) Hubb. is native to Africa, but is now widely naturalized throughout the tropics. It was first recorded from the Pacific Islands in 1903 (Hawai'i), where it is common in dry, disturbed areas, such as croplands and empty lots at up to 1900 m elevation, and is especially abundant in Hawai'i.

This grass can be distinguished by its glaucous foliage and loose panicles of ovate spikelets densely covered with pink to purple hairs that fade to silver. Synonyms: *Tricholaena repens* (Willd.) Hitchc., *T. rosea* Nees.

SACCHARUM SPONTANEUM
Poaceae (Grass Family)

COMMON NAMES: wild cane
DISTRIBUTION: Hawai'i, Guam, Belau

Perennial grass spreading by stout rhizomes. **Culms** erect, 2—4 m or more in height. **Leaf sheath** glabrous, with overlapping margins; ligule membranous, 3—4 mm long, with a rounded apex fringed with scattered hairs and a dense row of hairs 2—3 mm long behind it. **Leaf blade** mostly 50—90 x 5—15 cm. **Inflorescence** a plumose panicle 25—60 cm long with numerous racemes 3—15 cm long, rachis and pedicels hirsute, rachis falling off in units comprising a pair of spikelets and the rachis below them. **Spikelets** lanceolate, 3—4 mm long, bearded at the base, the hairs silky, 2—3 times as long as the spikelet, in unequally stalked pairs. Glumes equal, as long as spikelet, glabrous on the back. Upper lemma nearly as long as spikelet, membranous.

Saccharum spontaneum L. is native to tropical and temperate regions of Asia, and was probably first recorded from the Pacific Islands in 1903 (Hawai'i). It is common to abundant in disturbed lowland areas, often along the margins of wetlands, at up to 460 m elevation. It is particularly abundant in Guam, reportedly introduced there from Saipan, where it was grown experimentally by sugar companies. It tends to form dense stands that choke out all other vegetation, but is found in only a few places in Hawai'i, and is not reported from elsewhere in Polynesia.

This robust grass can be distinguished by its large, plumose inflorescence that breaks up into units bearing two spikelets, and peduncles hairy below the panicle. It is similar to sugar cane, which has thicker stems and a glabrous peduncle.

SETARIA VERTICILLATA
Poaceae (Grass Family)

COMMON NAMES: bristly foxtail; mau'u pilipili (Hawai'i)
DISTRIBUTION: Hawai'i, Guam, and the Marquesas

Tufted annual grass. **Culms** mostly 30—100 cm in height, often geniculate and branching at lower nodes, mostly pubescent. **Leaf sheath** compressed, glabrous or with ciliate margins; ligule a row of hairs 0.8—1.5 mm long. **Leaf blade** 4—10 mm wide, glabrous to sparsely pilose, margins scabrous. **Inflorescence** a cylindrical green to purple-tinged, densely flowered panicle 2—10 cm long, often doubling over on itself; branches short, whorled, bearing several spikelets, each of which is subtended by a bristle 2—6 mm long; axis finely hispid. **Spikelets** ovate, awnless, 1.2—1.8 mm long; glumes unequal, lower one *ca.* 1/3 as long as spikelet, upper nearly as long as spikelet, 5—7-nerved; first lemma as long as spikelet, 7-nerved.

Setaria verticillata (L.) P. Beauv. is native to Europe and was first recorded from the Pacific Islands in 1895 (Hawai'i). It is occasional to common in dry, disturbed areas, often in disturbed lowland forests on all the main islands of Hawai'i and in Guam, at up to 820 m elevation.

This grass can be distinguished by its green to purplish, dense cylindrical panicle sometimes doubling over on itself, spikelets borne several to each short branch, and single persistent bristle subtending the spikelet. Two similar species, *Setaria parviflora* (Poiret) Kerg. (synonym *S. gracilis* Kunth) in Hawai'i and *Setaria pumila* (Poiret) Roem. & Schult. (synonym *S. glauca* (L.) P. Beauv.) in the South Pacific, differ in having a yellowish, cylindrical panicle with several bristles below each spikelet.

SORGHUM SUDANENSE
Poaceae (Grass Family)

COMMON NAMES: Sudan sorghum
DISTRIBUTION: all the main island groups (a new record for Hawai'i)

Robust tufted annual grass. **Culms** mostly 1—2 m high, stout, erect, glabrous. **Leaf sheath** glabrous, often hairy at base; ligule membranous, 1—1.5 mm long with a fringe of hairs of similar length. **Leaf blade** 20—60 x 1—2.5 cm, surfaces glabrous, with a prominent yellow midrib, foliage and stems often spotted purple; margins scabrous. **Inflorescence** a panicle 15—35 cm long, hairy at the base, with many whorled, erect to ascending branches 4—15 cm long. **Spikelets** paired or in 3s, one sessile and bisexual, and 1 or 2 pedicellate male ones. Sessile spikelet lanceolate, 5—7 mm long, lower and upper glumes as long as the spikelet, silky-hairy; sterile lemma usually with a bent awn 5—15 mm long. Pedicellate spikelets lanceolate, 5—7 mm long.

Sorghum sudanense (Piper) Stapf is possibly native to North Africa and was first recorded from the Pacific Islands in 1895 (Samoa). It is occasional to common in disturbed places in the lowlands, especially on roadsides, and is often misidentified as *Sorghum halepense*.

This robust grass can be distinguished by its lack of rhizomes, terminal panicles, and spikelets in 2s or 3s, with one spikelet ovate, pubescent, often awned, sessile, and 5—7 mm long, and 1 or 2 lanceolate, awnless, stalked male spikelets of similar length. Synonyms: *Andropogon halepensis* of many authors, *Sorghum halepense, S. vulgare* Pers. var. *sudanensis* (Piper) Hitchc. The nearly identical *Sorghum halepense* (L.) Pers. of Hawai'i and elsewhere is a perennial with creeping rhizomes and slightly smaller spikelets.

SPOROBOLUS DIANDER
Poaceae (Grass Family)

COMMON NAMES: Indian dropseed
DISTRIBUTION: all the main island groups

Tufted perennial grass. **Culms** slender, erect, geniculate at the base, 20—60 cm long, nodes glabrous, leaves crowded at the base. **Leaf sheath** striate, glabrous except for the margins; ligule a ridge of hairs less than 0.2 mm long. **Leaf blade** linear, 4—20 x 0.2—0.4 cm; surfaces mostly glabrous; margins rolled or flat. **Inflorescence** a loose, narrow panicle 10—25 cm long, with erect to spreading branches 0.5—3 cm long. **Spikelets** laterally flattened, 1.3—1.6 mm long, pale green, 1-flowered, the glumes persistent on the pedicel when the grain falls. Glumes and lemma membranous, lower glume 1/4 as long as the spikelet, upper glume *ca.* 1/2 as long, lemma as long. Stamens 2. Grain obovate, 0.6—0.8 mm long, flattened, brown.

Sporobolus diander (Retz.) P. Beauv. is native to India and was first recorded from the Pacific Islands in 1911 (Hawai'i). It is common in lawns, roadsides, and other disturbed places at up to 460 m elevation.

This grass can be distinguished by its narrow leaves often with in-rolled margins, narrow panicles with short, spreading to erect branches, and ovate spikelets that persist on the rachis after the brown seed has fallen out. It is very similar to, hard to distinguish from, and often confused with the widespread *Sporobolus fertilis* (Steud.) Clayton, which differs in having its branches appressed to the axis at maturity. The taxonomy of this genus is not clear.

GLOSSARY OF BOTANICAL TERMS

Achene— A small, dry, non-splitting, 1-seeded fruit with the seed fused to the ovary wall, as in the grass family.

Acuminate— Referring to an acute apex that tapers to a long, drawn-out point, usually with somewhat concave sides.

Acute— Tapering to a sharp, but not drawn-out, point. Compare attenuate.

Alternate— Said of leaves arranged one per node. Compare opposite.

Annual— Having a life span of a single year or season. Compare perennial.

Anther— The pollen sac attached to a stamen. See filament.

Anthesis— The flowering period, or when the flower opens.

Apetalous— Lacking petals, said of flowers that have no corolla. In such cases, the sepals sometimes look like petals. See petaloid.

Apically— Towards the apex or tip of a structure.

Apiculate— Referring to an apex with a small, short, sharp flexible point.

Appressed— Lying flat against a surface, as appressed hairs.

Ascending— Rising or curving upward, as of stems.

Attenuate— Tapering gradually to form a long, straight-sided tip. Compare acute.

Awn— A slender, bristle-like appendage usually at the tip of a structure, especially on grass spikelets.

Axil— The upper angle between a leaf petiole and stem. Flowers situated in the axil are referred to as being axillary.

Bearded— Bearing a tuft of hairs.

Berry— A fleshy, pulpy fruit containing two or more seeds, such as a grape, passionfruit, or guava.

Bilabiate— Two-lipped, said of a corolla or calyx with the parts fused into an upper and lower lip.

Bipinnately compound— Twice pinnate, the first divisions being further divided into leaflets.

Bipinnatifid— Bipinnate with the leaflets lobed.

Bisexual— Having functional male and female organs in the same flower.

Bract— A reduced or modified leaf subtending a flower or inflorescence. A secondary bract is sometimes called a bracteole.

Bracteole— A secondary bract, often on the inflorescence but not directly below the flower.

Bulbil— A small bulb produced vegetatively in the leaf axil or sometimes on the margin.

Caducous— Falling off early, referring to sepals or petals.

Calyx— The outer, usually green whorl of the flower, enclosing the flower bud. It is composed of free or fused sepals. The plural is calyces.

Campanulate— Bell-shaped, said of a corolla or calyx.
Canescent— Gray pubescent or hoary.
Capsule—A dry, splitting fruit with several cells, opening by sections called valves.
Caudate— Having a long, narrow, tail-like apex.
Chaffy— Bearing thin, membranous scales in the inflorescence, like in the heads of some members of the sunflower family.
Ciliate— With a fringe of hairs on a margin (e.g., leaf margin).
Circumscissile— Describing a fruit whose top splits off along a seam.
Clavate— Club-shaped, referring to fruits or other structures.
Composite— Composed of two types of florets, said of the flowers of the sunflower family, Asteraceae (Compositae). See disc and ray floret.
Compound— Said of leaves with the blade further divided into leaflets or pinnae. Compare simple.
Compressed— Flattened in one plane.
Cordate— Heart-shaped, said of leaves.
Coriaceous— Having a leathery texture, said of leaves.
Corolla—The inner, usually colored, whorl of a flower, composed of free or fused petals, or sometimes absent.
Corona—A row of appendages between the stamens and petals.
Corymb— A flat-topped, short, broad inflorescence with the center flower the youngest. Corymbose is the adjective form.
Crenate— Round-toothed or scalloped, as of leaf margins.
Culm— The stem of a grass or sedge.
Cuneate— Wedge-shaped or triangular, referring to leaf bases.
Cuspidate— An apex (e.g., of a leaf) with a long, sharp-pointed tip with concave sides.
Cyathium— Type of inflorescence in spurge genera (*Euphorbia* and *Chamaesyce*) in which the unisexual flowers are clustered together within a bract-like envelope. The plural is cyathia.
Cyme—A cluster of flowers with the oldest ones at the end or center. Cymose is the adjective form. Compare panicle.

Decumbent— Prostrate, but with ascending stem tips.
Decurrent— Referring to a leaf blade base that tapers down to a narrow wing that extends to the stem.
Dentate— With sharp teeth perpendicular to the margin (of a leaf).
Diadelphous— Bearing stamens in two bundles, especially in papilionaceous flowers that usually have a 9 plus 1 arrangement. Compare monadelphous.
Dioecious— Condition of plants with unisexual flowers in which the male and female flowers are on separate plants. Compare monoecious.

Digitate— Shaped or arranged like the palm of the hand.
Dimorphic— Having or occurring in two forms.
Disc floret— A central, tubular flower of a "composite" inflorescence of the sunflower family (Asteraceae or Compositae). See ray floret.
Distichous— Referring to leaves that are all arranged in the same plane.
Dorsally— Referring to the back or outer side of a structure.
Drupe— A fleshy fruit with a single seed enclosed in a hard shell, such as a mango or peach; stone fruit.

Elliptic— Shaped like an ellipse. An ellipsoid is a 3-dimensional figure shaped in outline like an ellipse, said of some fruits.
Emarginate— Referring to a leaf with a notch at the tip.
Entire— Having a continuous margin lacking teeth or lobes.
Epipetalous— Borne on the corolla or petals, said of stamens.
Even-pinnately compound— Pinnate with an even number of leaflets, i.e., without a single unpaired one at the apex. Compare odd-pinnately compound.
Exserted— Sticking out, said of stamens when they protrude from the corolla.

Fascicle— A condensed cluster, said of leaves or flowers.
Filament— The stalk of a stamen, bearing the anther.
Filiform— Thread-like or filamentous.
Floret— A small flower of members of the sunflower family (Asteraceae or Compositae). See disc and ray florets.
Follicle— A dry splitting fruit opening on only one side.
Funnelform— Funnel-shaped, said of corollas of flowers.
Fusiform— Spindle-shaped, narrowed at the ends and thickened in the middle, said of fruits or seeds.

Geniculate— Bent like a knee, said of awns and filaments.
Glabrous— Said of a surface lacking pubescence; hairless.
Glandular— Bearing sticky glands or hairs on the surface.
Glaucous— Covered with a bloom, or a white substance that rubs off.
Globose— Spherical in shape.
Glume— One of two small, sterile bracts that form the outer series of a grass spikelet, but is sometimes reduced or absent.

Herb— A non-woody plant, usually small. Compare shrub.
Hirsute— Bearing long, coarse hairs.
Hispid— Bearing stiff, short hairs or bristles.
Hypanthium— The cup-like receptacle of some flowers; it bears the sepals, petals, and stamens.

Imbricate— Overlapping like shingles on a roof, said of flower parts such as bracts.
Included— Referring to stamens that do not protrude from the corolla. Compare exserted.
Indehiscent— Not splitting open, said of fruits.
Inferior— Said of an ovary or fruit that has the sepals on top, i.e., the ovary is inferior to the attachment of the sepals. Compare superior.
Inflorescence— A flower cluster or the arrangement of flowers on a plant.
Internode— The part of a stem between the nodes or leaf attachments.
Interpetiolar— Situated on a stem between two petioles, said of stipules.
Involucre— A whorl of leaves or bracts close to the base of a flower cluster.

Lanceolate— Lance-shaped in outline, several times longer than wide, with the widest portion towards the base of the leaf. Compare oblanceolate.
Laterally compressed— Compressed from side to side, rather than from top to bottom.
Leaflet— A division of a compound leaf.
Lemma— The scale within the glumes enclosing the flower of the grass family (Poaceae).
Ligule— A projection at the top of the leaf sheath in many grasses.
Limb— Expanded terminal portion of some corollas.
Linear— Long and narrow, with the sides almost parallel.
Lip— The lower or upper part of a corolla or calyx when the petals or sepals are unequal. See bilabiate.
Lyrate— Pinnately lobed with a large terminal lobe and smaller lateral ones.

Mericarp— A section of a schizocarp.
Monadelphous— Said of stamens united by their filaments into a single bundle. Compare diadelphous.
Monoecious— Condition of a plant with unisexual flowers when male and female flowers are on the same plant. Compare dioecious.
Mucro— A sharp, tooth-like tip of some leaves, bracts, petals, or other parts. Mucronate refers to a tip with a mucro.
Multiple fruit— Fruit formed from several flowers rather than one.

Nectary— A nectar-secreting gland on a leaf or flower.
Node— Point of attachment of a leaf on a stem.
Nerve— A distinct vein.
Nutlet— A small, 1-seeded, non-splitting lobe of a divided fruit.

Ob- — Prefix referring to a structure that is widest towards the tip rather than the base. See below.

Obconical— Shaped like a cone, with the broadest end at the tip.
Obcordate— Heart-shaped with the wider end towards the apex.
Oblanceolate— Lanceolate, but with the widest part towards the tip. Compare lanceolate.
Oblique— Unequally-sided, as in the bases of some leaves.
Oblong— Longer than broad, with the sides nearly parallel to each other.
Obovate— Ovate, but with the widest part towards the tip. Compare ovate.
Obovoid— Ovoid with the wider end towards the apex.
Obtuse— Blunt, rounded, as of a leaf tip.
Odd-pinnately compound— Pinnate with an odd number of leaflets, i.e., with an unpaired leaflet at the apex. Compare even-pinnately compound.
Opposite— Referring to leaves borne in pairs at the nodes. Compare alternate.
Orbicular— Round in outline. Compare suborbicular.
Ovary— The female part of the flower, containing the embryos or immature seeds.
Ovate— Oval in outline, with the widest part towards the base. Compare obovate.
Ovoid— Said of fruits, etc., that are oval in outline.

Palea— The membranous bract within the glumes of grass flowers that, along with the lemma, encloses the flower. Palaea is the plural.
Palmate— Lobed or divided in a hand-like fashion, usually in reference to leaf blades or their veins. Compare pinnate.
Panicle— A compound inflorescence with a main axis and racemose branches, with the youngest flowers towards the tip. Paniculate is the adjective form. Compare cyme.
Papilionaceous— Butterfly-like, said of a sweetpea type of flower of the legume family (Fabaceae or Leguminosae).
Pappus— A floral structure on the ovary of composite (Asteraceae) flowers, often modified into a plume, scale, or bristle.
Pedicel— The stalk of a flower. Pedicellate means borne on a pedicel.
Peduncle— The stalk of a flower cluster.
Peltate— Referring to a leaf that has the petiole joined to the blade away from or inside the margin.
Perennial— Living more than one season or year. Compare annual.
Persistent— Not falling, remaining on the plant or flower part.
Petal— A division of a corolla.
Petaloid— Petal-like in texture or color, said of the calyx of some flowers.
Petiole— The stalk of a leaf.
Pilose— Bearing soft, relatively long hairs.
Pinna— The first division of a compound leaf; the pinna is further divided into leaflets. Pinnae is the plural.
Pinnate— Divided in feather-like fashion. Compare palmate.

Pinnately compound— Describing a leaf formed from several leaflets arranged in feather-like fashion.

Pinnatifid— Parted or lobed in a feather-like fashion.

Pistillate— Referring to a unisexual flower with only female parts present or functional. Compare staminate.

Plumose— Feather-like, said of the pappus of some composite (Asteraceae) flowers.

Procumbent— Growing flat along the ground but not rooting at the nodes.

Puberulent— Finely pubescent.

Pubescent— Hairy, covered with soft hairs.

Punctate— Bearing translucent dots or glands on the surface.

Raceme— Simple, elongated inflorescence with stalked flowers on a single main axis (rachis), the youngest ones at the top. Racemose is the adjective form.

Rachis— The axis of a compound leaf or inflorescence. Rachises is the plural.

Ray— A spreading branch of an inflorescence, e.g. of an umbel.

Ray floret— A strap-shaped flower of a composite of inflorescence, usually in members of the sunflower family (Asteraceae or Compositae). Compare disc floret.

Recurved— Bent backward, often said of prickles with a short terminal hook.

Reflexed— Abruptly bent downward or backward.

Reniform— Kidney-shaped.

Revolute— Having the margins rolled towards the lower surface, said of leaves.

Rhizome— An underground stem, usually with nodes and buds. Rhizomatous means bearing rhizomes.

Rosette— With the leaves arranged in a dense spiral at the base of the plant.

Rotate— Wheel-shaped, as in rotate corollas.

Rugose— Having a wrinkled surface.

Sagittate— Shaped like an arrowhead.

Salverform— Said of a corolla with a slender tube and abruptly expanded limb.

Scabrid— Having a rough or finely serrate edge or surface.

Scabrous— Feeling rough, usually because of short stiff hairs or scales.

Scandent— Weakly climbing, but lacking tendrils.

Scale— Dry leaf or bract.

Scapose— Said of an inflorescence having a long stalk (scape).

Scarious— Dry and membranous, referring mostly to flower parts.

Schizocarp— A dry fruit splitting apart at maturity into 1-seeded segments called mericarps.

Scorpioid cyme— A cyme coiled like the tail of a scorpion.

Sedge— Member of the monocot family Cyperaceae.

Sepal— A division of the calyx.

Serrate— Having a saw-toothed margin, as of some leaves.
Sessile— Lacking a stalk, said of leaves, flowers, etc.
Sheath— Oblong tubular structure surrounding a plant part, such as a leaf sheath, which is the basal part of a grass leaf that surrounds the stem.
Shrub— A relatively short, woody plant with branches forming near to the ground.
Simple— Said of a leaf or other structure that is not divided into parts. Compare compound.
Sinus— Space between two lobes or divisions of a leaf.
Spathe— An often showy leaf or bract subtending a spadix inflorescence.
Spathulate— Spoon-shaped.
Spike— An unbranched inflorescence bearing sessile flowers, the youngest ones at the tip.
Spikelet— A grass or sedge inflorescence consisting of flowers and membranous scales.
Spinulose— Bearing small spines.
Spreading— Bending or growing outward or horizontally.
Stamen— The male part of the flower, consisting of an anther and a filament.
Staminate— Unisexual flower bearing stamens but no functioning female parts. Compare pistillate.
Staminode— A sterile stamen, lacking the anther.
Stellate— Star-shaped, said of some hairs.
Stigma— The sticky receptive tip of an ovary, with or without a style.
Stipules— Paired basal appendages present on the petioles of some plants.
Stolon— A horizontal stem rooting at the nodes or producing a new plant at its tip.
Striate— Finely grooved.
Strigose— Bearing sharp, straight hairs appressed to the surface; often the hairs have swollen bases.
Style— The stalk between the ovary and stigma.
Sub- — Prefix meaning less than or nearly. See entries below.
Subglobose— Nearly spherical in shape.
Suborbicular— Nearly round in outline. Compare orbicular.
Subretuse— Faintly or nearly notched (retuse) at the apex.
Subsessile— Nearly stalkless or sessile.
Subshrub— Intermediate between a shrub and an herb.
Subtend— Attached below something.
Subulate— Awl-shaped.
Succulent— With fleshy stems or leaves.
Superior— Said of an ovary or fruit with the sepals on the bottom. Compare inferior.
Syconium— Kind of flower or fruit of the genus *Ficus* in which the receptacle forms a globose structure with the flowers inside.
Sympetalous— Having the petals fused together to form the corolla.

Taproot— A thickened, usually solitary root of some plants.
Tendril— A thread-like plant organ that by its rotating growth allows a plant to become attached to another plant for support.
Terminal— Situated at the end of a branch or rachis.
Throat— The part of a sympetalous corolla where the limb joins the tube.
Tomentose— Densely woolly with tangled, matted hairs.
Toothed— Bearing teeth or indentations along the margin, said of leaf margins.
Tree— A woody plant with a single trunk and lacking branches near the base. Compare shrub.
Trifoliate— Bearing leaves divided into three leaflets.
Truncate— Appearing cut-off or squared at the end, as of a leaf tip.
Tube— Basal, cylindrical portion of a corolla having fused petals.
Tuber— An underground, swollen, root-like stem of some plants.
Tubercle— A rounded protruding body. Tuberculate means covered with tubercles.
Tufted— Growing in clumps.
Turbinate— Top-shaped, referring to a fruit or seed.

Umbel— A flat- or round-topped inflorescence with the stalks of the flowers all arising from one point, the oldest flowers at the center. Umbellate is the adjective form, and umbelliform means "similar to an umbel."
Unisexual— Said of flowers lacking either stamens or an ovary.
Urceolate— Urn-shaped, usually referring to a corolla.
Utricle— A bladder-like, 1-seeded fruit of some plants.

Valve— A section or piece into which a capsular fruit splits. Valvate refers to structures bearing valves.
Villous— Shaggy with long, soft hairs.

Whorled— Said of leaf arrangements having more than two leaves per node.
Winged— Having ridges, said of some stems with longitudinal ridges running down the stem or petiole.

INDEX TO SPECIES

The 170 species in **bold** are featured ones, those in regular print are only mentioned in the text, and those in *italics* are synonyms. The capital letters following some of the species names are the wetland status in the Hawaiian region (Reed 1988): OBL= obligate wetland species; FACW = facultative wetland species; FAC= facultative species; FACU = facultative upland species; *, +, and - indicate degree within the category. Species names without any listing are considered to be upland species (UPL) that are rarely if ever found in wetlands. A listing of NI indicates that insufficient information was available ("no indicator").

INDEX TO COMMON NAMES

This index includes the common English names of the plants found in this book. The most common local names are also included, along with their origin (H for Hawai'i, G for Guam).

SELECTED BIBLIOGRAPHY

Christophersen, E. 1935, 1938. Flowering plants of Samoa. B. P. Bishop Museum Bull. 128: 1—221; II. 154: 1—77.

Fosberg, F. R., M.-H. Sachet, and R. Oliver. 1979. A geographical checklist of the Micronesian Dicotyledonae. Micronesica 15: 1—295.

Fosberg, F. R., M.-H. Sachet, and R. Oliver. 1982. A geographical checklist of the Micronesian Monocotyledonae. Micronesica 18(1): 23—82.

Merlin, M. D. 1976. Hawaiian forest plants. Oriental Publishing Co., Honolulu. 68 pp.

Merlin, M. D. 1977. Hawaiian coastal plants and scenic shorelines. Oriental Publishing Co., Honolulu. 68 pp.

Raulerson, L. and A. Rinehart. 1991. Trees and shrubs of the Northern Mariana Islands. Coastal Resourses Management Office of the Governor, Saipan. 120 pp.

Reed, P. B. Jr. 1988. National list of plant species that occur in wetlands: Hawaii (Region H). U. S. Fish and Wildlife Service, Washington, DC. 88 pp.

Smith, A. C. 1979—1991. Flora Vitiensis nova; a new flora of Fiji. National Tropical Botanical Garden. 5 vols.

Stemmermann, L. 1981. A guide to Pacific wetland plants. U. S. Army Corps of Engineers, Honolulu District. 118 pp.

Stone, B. 1970. Flora of Guam. Micronesica 6: 1—659.

Sykes, W. R. 1970. Contributions to the flora of Niue. New Zealand Dept. Sci. & Indust. Res. Bull. 200: 1—321.

Wagner, W. L., D.R. Herbst, and S. H. Sohmer. 1990. Manual of the flowering plants of Hawai'i. University of Hawaii Press & Bishop Museum Press, Honolulu. 2 vols.

Whistler, W. A. 1983. Weed handbook of western Polynesia. Deutsche Gesellschaft für Technische Zusammenarbeit, Eschborn, Germany. 151 pp.

Whistler, W. A. 1988. Checklist of the weed flora of western Polynesia. Technical Paper 194, South Pacific Commission, Noumea. 69 pp.

Whistler, W. A. 1992a. Flowers of the Pacific Island seashore. Isle Botanica, Honolulu. 154 pp.

Whistler, W. A. 1992b. Polynesian herbal medicine. National Tropical Botanical Garden, Lawai, Kaua'i. 238 pp.

Yuncker, T. G. 1959. Plants of Tonga. B. P. Bishop Museum Bull. 220: 1—343.